Neurophysiological and Clinical Aspects of Vestibular Disorders

Advances in
Oto-Rhino-Laryngology

Vol. 30

Series Editor
C. R. Pfaltz, Basel

S. Karger · Basel · München · Paris · London · New York · Tokyo · Sydney

Neurophysiological and Clinical Aspects of Vestibular Disorders

Volume Editor
C. R. Pfaltz, Basel

134 figures and 39 tables, 1983

KARGER

S. Karger · Basel · München · Paris · London · New York · Tokyo · Sydney

Advances in Oto-Rhino-Laryngology

National Library of Medicine, Cataloging in Publication
  Bárány Society. Extraordinary Meeting
  (8th: 1982: Basel, Switzerland)
  Neurophysiological and clinical aspects of vestibular disorders/
  8th Extraordinary Meeting of the Bárány Society, Basel, June 22-25, 1982;
  Volume editor, C. R. Pfaltz. – Basel; New York: Karger, 1983.
  (Advances in oto-rhino-laryngology; v. 30)
  1. Labyrinth Diseases – physiopathology – congresses 2. Vestibular Apparatus – physiopathology – congresses
  I. Pfaltz, C. R. (Carl Rudolf) II. Title III. Series
  W1 AD701 v. 30 [WV 255 B2255 1982n]
  ISBN 3-8055-3607-0

Drug Dosage
  The author and publisher have exerted every effort to ensure that drug selection and dosage set forth in this text are in accord with current recommendations and practice at the time of publication. However, in view of ongoing research, changes in government regulations, and the constant flow of information relating to drug therapy and drug reactions, the reader is urged to check the package insert for each drug for any change in indications and dosage and for added warnings and precautions. This is particularly important when the recommended agent is a new and/or infrequently employed drug.

Contents

Neurophysiological and Clinical Aspects of the Vestibulo-Ocular and the Optokinetic Reflex

Contents

Vestibular Tests and Diagnosis of Intracranial Pathology by Neuro-Otological Approaches

Visual-Vestibular Interaction

Experimental and Clinical Aspects of Ménière's Disease and Other Vestibular Disorders

Ataxia

Neurophysiological and Clinical Aspects of Vestibular Compensation

Medical Treatment of Vertigo. How to Evaluate Its Effect?

Preface

The Bárány Society is an interdisciplinary scientific panel founded in 1960 on the initiative of the late *C.S. Hallpike* and *C.O. Nylen* to the memory of *Robert Bárány*, who was awarded the Nobel prize in 1917 for his outstanding research in the field of neuro-otology. The founders of the Bárány Society aimed at increasing contacts between scientists in vestibular research because at that time relatively few basic scientists and clinicians took an interest in this particular field of neuroscience. Since the beginning of space exploration medical problems concerned with space and weightlessness and hence the interest in the physiology and pathophysiology of the vestibular system have become more and more important. Vestibular research is no longer a restricted area of otology, covered by experienced clinicians with a scientific interest in the function of the semicircular canals, but has become a part of the vast interdisciplinary field of central nervous system data processing research. In up-to-date vestibular research emphasis is laid on the vestibulo-ocular and the optokinetic reflex mechanisms, because 'rapid progress has been made in recent years toward understanding how the C.N.S. processes visual and vestibular signals to produce eye movements and body postural responses ... and a new literature on visual-vestibular interactions is beginning to build ...' [*Cohen, 1981*]. This evolution demands and implies an increasing contribution from basic sciences. For these reasons topics such as visual correlates of full-field motion, natural retinal image motion, the influence of visual motion cues on postural control, input-output activity of the primate flocculus during visual-vestibular interaction, error signals subserving adaptive gain control in the primate vestibulo-ocular reflex have become fundamental problems which demand the cooperation of experts like the neurophysiologist, ophthalmologist, neurologist, psychologist and otologist. At present it is practically impossible to follow or digest the vast information produced in the laboratories of basic science institutions and of aeronautic as well as space medicine. This situation, however, implies the danger that the clinician is no longer informed about recent advances in vestibular research which are relevant to his understanding of clinical symptoms of a vestibular disorder. On the other hand, there is also a certain danger that basic research gradually follows its own line without having any concern with problems of clinical relevance, becoming more and more a scientific 'l'art pour l'art' activity.

Planning the 8th extraordinary meeting of the Bárány Society in Basel, we had in mind to organize a symposium on 'neurophysiological and clinical aspects of vestibular disorders' because we wanted an interdisciplinary discussion not only between the representatives of various medical specialties but also principally between the open-minded scientist from the laboratory and the scientifically minded clinician. For these reasons main topics of general and mutual interest have been chosen and they were elucidated during the meeting both from a theoretical and clinical angle. The present volume is an attempt to give a state of the art review on some interdisciplinary topical problems in experimental and clinical vestibular research.

Basel, July 1982

C.R. Pfaltz

Acknowledgements

The committee of the Bárány Society would like to thank the government and the University of Basel for taking over the official patronage of our symposium. Special thanks go to the management of Ciba-Geigy Ltd., Basel, for providing major support for the organization of this meeting.

Adv. Oto-Rhino-Laryng., vol. 30, pp. 1–8 (Karger, Basel 1983)

Habituation and Plasticity of the Vestibulo-Ocular Reflex

Volker Henn

Neurological Clinic, University Hospital, Zürich, Switzerland

The classical vestibulo-ocular reflex (VOR) is composed of three neurons, yet normal nystagmus in response to vestibular stimulation involves neuronal activity in many brainstem areas [16]. Therefore, this reflex is very reliably elicited, but also highly modifiable. There have been many attempts to classify the different mechanisms which can modify the VOR. Strictly as operational terms, I will use the expressions habituation, plastic changes, and recovery after lesions [12].

Habituation refers to a response decline which persists over time, up to months or years. This differentiates it from fatigue. Habituation of the VOR occurs while the animal or human subject is exposed repeatedly to the same stimulus pattern in darkness (review [7]). Effective are velocity steps or low frequency sinusoidal stimulation, whereas high frequency sinusoidal stimulation has proved to be ineffective [13]. Figure 1 shows an example from a monkey in which the time constant of the slow phase of nystagmus to a velocity step has decreased from 51 to 14 s within 1 h. During the hour, between the two step stimuli, the monkey was continuously rotated with a sinusoidal velocity profile at a frequency of 0.01 Hz. This led to an increasing phase angle of slow phase eye velocity relative to turntable velocity. An increasing phase angle is equivalent to a shorter time constant. Simultaneously the gain, i.,e. the maximum eye velocity relative to maximum turntable velocity, has decreased. There is always a good correspondence between time con-

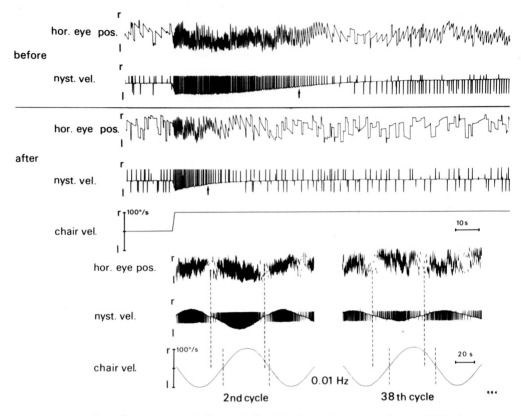

Fig. 1. Step responses before and after 64 min continuous rotation at 0.01 Hz in a monkey (from [13]).

stants measured after velocity steps and time constants calculated from frequency response curves. The increasing shift between occurrence of maximum eye velocity and turntable velocity at lower frequencies as well as the decreasing maximum eye velocity can be plotted in a Bode diagram (fig. 2). In general, it was found that only frequencies of stimulation at which some phase advance was already evident were effective in producing habituation. These include frequencies below 0.02 Hz. If the animal was habituated only with one frequency, e.g. 0.01 Hz, the whole frequency curve shifted as shown in figure 2. Even a frequency of 0.002 Hz, at which the naive monkey already showed a phase advance of 90°, still changed the whole frequency curve in the medium

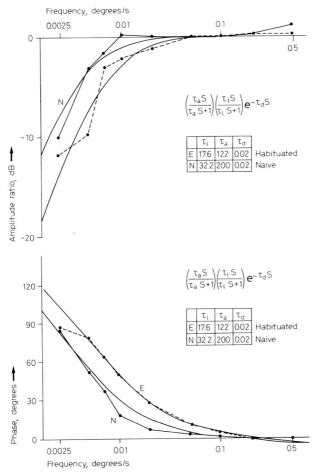

Fig. 2. Frequency response curve from averaged VOR responses in monkeys together with adaptation model results. Measurement points were connected by straight line segments, continuous segments average values from 5 naive monkeys, dotted segments from 8 habituated monkeys. Superimposed are model calculations with values indicated in the insert (from [5]).

range, as in other animals in which the whole spectrum of frequencies had been applied.

Humans can be habituated in a similar way [14]. In a recent study 16 normal subjects were exposed to sinusoidal velocity stimulation at frequencies between 0.005 and 0.01 Hz [1]. In all 16 subjects, after only

½ h of rotation, an increase in low frequency phase advance was seen together with a decrease of the nystagmus time constant after velocity steps. On average, time constants declined from 20 to 13 s. However, retention of that shortened time constant was seen only in one third of the subjects. Whereas in monkeys retention of a shortened time constant was observed from day to day or even for several weeks or months, in many humans time constants returned to original values within 24 h. It is not known why some human subjects show such a good retention of the short time constant and some others do not. In addition, it is also not known what stimulus configuration leads to a lengthening of the time constant. No stimulus pattern has been described so far which would be suitable to prolong time constants under controlled laboratory conditions.

As time constants can be changed in normal human subjects and as they show different patterns of retention, the question has to be considered what are normal values. This is especially important in the evaluation of frequency response curves when patients with presumed vestibular disorders are investigated. When values fall outside certain standard limits, it must be determined whether subjects had been previously habituated, in contrast to a change in their time constants due to repeated diagnostic testing, or the progression of a disease process.

Transfer of habituation has been a matter of controversy. However, if stimulus patterns are such that they lead to a similar sensation, habituation can occur, e.g. habituation from optokinetic to vestibular nystagmus [24]. Several investigations are under way to explore to what extent such transfer of habituation could be used to the advantage of the patient in whom the aim is to restore normal nystagmus responses after partial lesions [18, 20].

Single neuron recordings in monkeys have shown that the time constant of activity in the vestibular nerve is around 5 s [9]. In the vestibular nuclei this signal is partially integrated which leads to a larger time constant, with a value between 8 and 30 s, and even up to 100 s in experimentally naive monkeys [4]. This integration process has been described as velocity storage [22] and is a decisive element in all models of the vestibular system, especially models describing visual-vestibular interaction (review [12]). All central vestibular neurons have time constants which are similar to each other and similar to that of accompanying vestibular nystagmus. During the process of habituation these values covary.

Plastic changes refer to an altered response after the subject's exposure to a novel combination of stimuli. *Gonshor and Melvill Jones* [10] introduced this kind of experiment by wearing prism spectacles which lead to a left-right reversal of the visual world. After wearing such prisms for several days and weeks, the VOR, measured in total darkness, first decreased and finally reversed. The VOR testing was done by rotating subjects on a turntable with a sinusoidal velocity profile, which is a predictable stimulus. As these experiments have been further refined, *Melvill Jones and Gonshor* [17] now interpret their results as consisting of two elements. A simple one leads to a long-lasting gain attenuation, and a complex one probably contains a predictive element, which enables the subject to fully adapt to the novel stimulus condition.

Many more experiments in humans as well as in animals have been performed using other forms of changing the visual input: magnifying and telescopic lenses, or other distortions of the visual image. All these experiments show the plasticity of the VOR over wide ranges, and the geometric specificity of the induced changes [2]. Changing the direction of the VOR has been successfully performed only in one experiment so far by *Schultheiss and Robinson* [23]. Although the flocculus seems to be an important interaction site for visual-vestibular interaction, extensive single neuron studies in monkeys could not identify a single site as being responsible for plastic changes [19].

Recovery after lesions refers to changes which gradually lead to a normalization of a previously pathologically altered vestibular response. *Precht* [21] and coworkers have shown for the horizontal system that the commissural fibers between the vestibular nuclei are important in that neurons in the vestibular nuclei regain some of their spontaneous activity and sensitivity to acceleration after unilateral labyrinthine lesions. Output from the vestibular nuclei, stronlgy asymmetric after such a lesion, is balanced again. This process of balancing does not need the flocculus [11, 15].

After labyrinthine lesions the gain of the VOR is usually depressed. Under normal natural conditions vestibular reflexes work in synergy with neck and optokinetic reflexes. Therefore, the hypothesis had to be checked whether after partial lesions such other reflexes could compensate the loss. The neck reflex or cervico-ocular reflex (COR) is tested by rotating the body with the head held stationary in space. In humans, responses are variable and have little gain. In monkeys this reflex is virtually absent. However, after labyrinthine lesions, the COR gain can increase contributing to restore compensatory eye movements [8].

In recent years the pathophysiology of the vestibular system has been explored by subjecting animals to partial vestibular lesions [6]. This can be done by labyrinthectomy, cutting of individual nerve branches of the vestibular nerve, or by plugging individual canals. With such a canal plugging operation, the build-up of pressure gradients across the cupula is made impossible while resting discharge in the nerve is supposed to remain normal. In such animals, within a short time, the COR becomes visible, and its gain rises over several weeks. In addition, a possible otolith contribution can be seen if the head is rotated over a stationary trunk. This leads to effectively compensatory eye movements in the high frequency range using high accelerations. There are profound differences between the recovery or compensatory gain increases of reflexes after canal plugging operations and cutting of the nerve. After lesions of the nerve: optokinetic nystagmus is reduced in the high velocity range, optokinetic afternystagmus is absent, during off-vertical axis rotation (barbecue spit rotation) only a peak-to-peak modulation is seen without a bias component, and CORs do not increase to achieve full compensation. If canals are plugged: optokinetic nystagmus and afternystagmus are essentially normal, during off-vertical axis rotation normal responses can be elicited, and it also seems that the gain of the COR can reach higher values [3]. These experiments show that a well localized peripheral lesion does not only lead to the loss of specific function, but can also affect other central functions in a profound way. In this particular instance it is the integrity of central vestibular cells which require the high tonic input from the nerve to function normally. It has not been clinically explored so far, to what extent this new knowledge could be used to differentiate between pure labryinthine, nerve or central lesions.

There are several different mechanisms which can modify the VOR. Some of them, like habituation, operate under normal conditions; others appear only if novel combinations of stimuli are brought into play. All subjects who have to correct vision by wearing spectacles experience such plastic changes, as this always leads to a changed input of the visual image. As the physiology and pathophysiology of these processes are investigated and a better understanding is achieved, one can be optimistic that some of this knowledge will also find its way into the clinic for diagnostic and therapeutic purposes.

References

1   Baloh, R.W.; Henn, V.; Jaeger, J.: Habituation of the human vestibulo-ocular reflex by low frequency harmonic acceleration. Am. J. Otolaryngol. *3:* 235–241 (1982).
2   Berthoz, A.; Melvill Jones, G.; Bégué, A.M.: Differential visual adaptation of vertical canal-dependent vestibulo-ocular reflexes. Exp. Brain Res. *44:* 19–26 (1981).
3   Böhmer, A.; Henn, V.; Suzuki, J.: Compensatory eye movements in the monkey during high frequency sinusoidal rotations; in Roucoux, Crommelinck, Physiological and pathological aspects of eye movements, pp. 127–130 (Junk, The Haag 1982).
4   Buettner, U.W.; Büttner, U.; Henn, V.: Transfer characteristics of neurons in vestibular nuclei of the alert monkey. J. Neurophysiol. *41:* 1614–1628 (1978).
5   Buettner, U.W.; Henn, V.; Young, L.R.: Frequency response of the vestibulo-ocular reflex (VOR) in the monkey. Aviat. Space Environ. Med. *52:* 73–77 (1981).
6   Cohen, B.; Suzuki, J.; Raphan, T.; Matsuo, V.; de Jong, V.: Selective labyrinthine lesions and nystagmus induced by rotation about off-vertical axes; in Lennerstrand, Zee, Keller, Functional basis of ocular motility disorders, pp. 337–346 (Pergamon, Oxford 1982).
7   Collins, W.E.: Habituation of vestibular responses with and without visual stimulation; in Kornhuber, Vestibular system. Handbook of sensory physiology, vol. VI/2, pp. 369–388 (Springer, Berlin 1974).
8   Dichgans, J.; Bizzi, E.; Morasso, P.; Tagliasco, V.: Mechanisms underlying recovery of eye-head coordination following bilateral labyrinthectomy in monkeys. Exp. Brain Res. *18:* 548–562 (1973).
9   Fernandez, C.; Goldberg, J.M.: Physiology of peripheral neurons innervating semicircular canals of the squirrel monkey. II. Response to sinusoidal stimulation and dynamics of peripheral vestibular system. J. Neurophysiol. *34:* 661–675 (1971).
10 Gonshor, A.; Melvill Jones, G.: Extreme vestibulo-ocular response induced by vision-reversal during head rotation. J. Physiol., Lond. *256:* 381–414 (1976).
11 Haddad, G.M.; Friendlich, A.R.; Robinson, D.A.: Compensation of nystagmus after VIIIth nerve lesions in vestibulo-cerebellectomized cats. Brain Res. *135:* 192–196 (1977).
12 Henn, V.; Cohen, B.; Young, L.R.: Visual-vestibular interaction in motion perception and the generation of nystagmus. Neurosci. Res. Prog. Bull. *18:* 457–651 (1980).
13 Jäger, J.; Henn, V.: Habituation of the vestibulo-ocular reflex (VOR) in the monkey during sinusoidal rotation in the dark. Exp. Brain Res. *41:* 108–114 (1981a).
14 Jäger, J.; Henn, V.: Vestibular habituation in man and monkey during sinusoidal rotation. Proc. N.Y. Acad. Sci. *374:* 330–339 (1981b).
15 Jeannerod, M.; Courjon, J.H.; Flandrin, J.M.; Schmid, R.: Supravestibular control of vestibular compensation after hemilabyrinthectomy in the cat; in Flohr, Precht, Lesion-induced neuronal plasticity in sensorimotor systems, pp. 208–220 (Springer, Berlin 1981).
16 Lorente de Nó, R.: Vestibulo-ocular reflex arc. Archs Neurol. Psychiat., Chicago *30:* 245–291 (1933).
17 Melvill Jones, G.; Gonshor, A.: Oculomotor response to rapid head oscillation

(0.5–5.0 Hz) after prolonged adaptation to vision-reversal. Simple and complex effects. Exp. Brain Res. *45:* 45–58 (1982).

18 Meran, A.; Pfaltz, C.R.: Der akute Vestibularisausfall. Akt. Neurol. *6:* 27–38 (1979).

19 Miles, F.A.; Lisberger, S.G.: Plasticity in the vestibulo-ocular reflex: a new hypothesis. Annu. Rev. Neurosci. *4:* 273–299 (1981).

20 Pfaltz, C.R.; Kato, I.: Vestibular habituation – interaction of visual and vestibular stimuli. Archs Otolar. *100:* 444–448 (1974).

21 Precht, W.: Characteristics of vestibular neurons after acute and chronic labyrinthine destruction; in Kornhuber, Vestibular system. Handbook of sensory physiology, vol. VI, pp. 451–462 (Springer, Berlin 1974).

22 Raphan, T.; Matsuo, V.; Cohen, B.: Velocity storage in the vestibulo-ocular reflex arc (VOR). Exp. Brain Res. *35:* 229–248 (1979).

23 Schultheiss, L.W.; Robinson, D.A.: Directional plasticity of the vestibulo-ocular reflex in the cat. Ann. N.Y. Acad. Sci. *374:* 504–512 (1981).

24 Young, L.R.; Henn, V.: Selective habituation of vestibular nystagmus by visual stimulation. Acta oto-lar. *77:* 159–166 (1974).

V. Henn, MD, Neurological Clinic, University Hospital,
CH-8091 Zürich (Switzerland)

Adv. Oto-Rhino-Laryng., vol. 30, pp. 9–12 (Karger, Basel 1983)

Age-Dependent Changes in the Gains of the Vestibulo-Ocular Reflex in Humans

Toshiaki Yagi[a], *Shuji Sekine*[a], *Motohiro Shimizu*[b]

[a] Department of Otolaryngology, Nippon Medical School, Tokyo, Japan;
[b] Department of Otolaryngology, Dokkyo Medical College, Tochigi, Japan

The function of the vestibulo-ocular reflex (VOR) is to maintain a stable retinal image during head rotation by generating appropriate compensatory eye movements. In the human, this reflex has a gain (measured as slow-phase eye velocity/head velocity, in the dark) of about 0.6 at a relatively low-frequency range. It has been reported [1, 7] that patients with cerebellar lesions, such as cerebellar atrophy, in which the number of cerebellar neurons are decreased, showed a high VOR gain as compared to normal subjects. On the other hand, it has been well known that the population of cerebellar neurons gradually decreases with age. Thus, it can be expected that aging probably influences the VOR gain control. Therefore, in this study, we examined the VOR gain under normal conditions and 1 h after wearing vision reversal prisms of subjects ranging in age from 5 to 79 years.

80 subjects (46 males and 34 females) who were free from vestibular and oculomotor disorders were used. They were divided into nine groups according to their ages, i.e., 5–9, 10–14, 15–19, 20s, 30s, 40s, 50s, 60s, and 70s, respectively. The subjects sat on the chair that was rotated sinusoidally in a horizontal plane at 0.25 Hz and 30° angular amplitude. Eye movement (E), eye velocity ($\dot{E}$) and head velocity ($\dot{H}$) were recorded. First, the subjects were asked to watch a small, dim light fixed approximately 2 m in front of the chair in the darkened room (baseline). Next, the subjects' eyes were covered by blackout goggles (initial level). In the youngest group, the examination was discontinued at this point, because of difficulty in obtaining cooperation for further experiments. After that, the subjects wore horizontal vision rever-

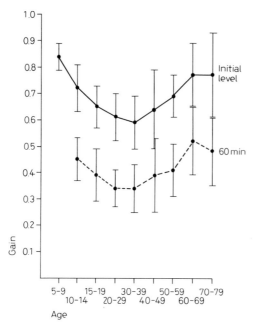

Fig. 1. Means (filled circles) and standard deviations (vertical bars) of the VOR gain of the initial level and 60 min after wearing vision reversal prisms.

sal prisms. During the image-reversal period, the subjects were asked to watch the dim light. Every 15 min, the subjects wore the blackout goggles and E, Ė and Ḣ were recorded in addition to the baseline and initial level recordings. The experiment was carried out in darkness, except for the small, dim target light.

Gain of the VOR was calculated for each subject as follows. The relative VOR gain (G) was estimated as the ratio of Ė in the baseline condition (Ėbl) to the eye velocity in the eye-covered condition (Ėx); G = Ė/Ėbl, since the VOR gain at this frequency is close to one in the baseline condition.

Means and standard deviations of the VOR gain in darkness before (initial level) and after wearing prisms (60 min) of each age group are shown in figure 1. Regarding the gain of the initial level, the gain was the highest in the youngest age group (mean: 0.84), gradually decreased with age and reached the lowest (mean: 0.59) in the 30s. Then the gain gradually increased again with age and reached 0.77 in the 70s. This means that the VOR gain in the dark is strongly influenced by age.

The adaptation of the VOR gain after 1 h of wearing vision reversal prisms is illustrated by the filled circles connected with dashed lines (fig. 1). As indicated in the figure, the amount of reduction of the VOR gain in each group was almost the same as that of the mean VOR gain of the initial level in each age group, so that the dashed line which represents the former, is nearly parallel with the solid line, which represents the latter. When we compared the net adaptation between each group, the gain of 1 h after wearing prisms divided by the gain of the initial level, no statistical difference was observed between them. Thus, it can be said that the adaptation phenomenon of the VOR gain control is not affected by aging.

The mean value of the VOR gain in the dark before wearing vision reversal prisms was lowest in the 30s, gradually increasing with age. This finding may be due to the reduction of the number of cerebellar Purkinje cells, which inhibit the activity of the vestibular neurons [2], with age, since it has been well known that the population of cerebellar Purkinje cells decreases with age, by approximately 2.5% per decade [3]. It has also been reported that, in some pathological conditions, the VOR showed an abnormally high gain [1, 7].

The VOR gain, however, also increased with decreasing age and reached the highest value in the youngest age group. We speculate that the high VOR gain in humans is innate, since in the higher mammals, such as monkeys [4] and cats [5], the VOR gain in the dark is close to one. Thus, the VOR gain in humans is probably the highest in newborns who have developed VOR arc and gradually decreased with age because the visual environment in human life is quite different from that of animals. In human life, the visual and vestibular inputs quite frequently conflict with each other so that the VOR produces plastic changes (learning) with age. Thus, we speculate that at younger ages, the learning process is predominant over the aging effect and that, at older ages, the aging effect is predominant over the learning process in the VOR gain control.

Recently, the plastic changes of the VOR have been intensively investigated in both humans and animals. In normal subjects, the VOR gain after 1 h of wearing vision reversal prisms became approximately 50–60% that of the initial level. However, in some neurological cases, the adaptation of the VOR is quite diminished in spite of the normal gain of the initial level reported by *Yagi* et al. Thus, it was speculated that the mechanisms of VOR gain control in darkness and the process of adaptation to vision reversal have different anatomical locations. And in the present study, we observed age-dependent changes in the VOR gain in darkness before wearing prisms and no change in the VOR adaptation after wearing prisms. This finding also supports the hypothesis mentioned above.

References

1   Baloh, R.W.; Jenkins, H.A.; Honrubia, V.; Yee, R.D.; Lau, C.G.Y.: Visual-vestibular interaction and cerebellar atrophy. Neurology 29: 116–119 (1979).
2   Fukuda, M.; Highstein, S.M.; Ito, M.: Cerebellar inhibitory control of the vestibulo-ocular reflex investigated in rabbit IIIrd nucleus. Exp. Brain Res. 14: 511–526 (1972).
3   Hall, T.C.; Miller, A.K.H.; Corellis, J.A.N.: Variations in the human Purkinje cell population according to age and sex. Neuropath. appl. Neurobiol. 1: 267–292 (1975).
4   Miles, F.A.; Fuller, J.H.: Adaptive plasticity in the vestibuloocular responses of the rhesus monkey. Brain Res. 72: 512–516 (1974).
5   Robinson, D.A.: Adaptation gain control of vestibuloocular reflex by the cerebellum. J. Neurophysiol. 39: 54–69 (1976).
6   Yagi, T.; Shimizu, M.; Sekine, S.; Kamio, T.: New neurotological test detecting cerebellar dysfunction. Vestibulo-ocular reflex changes with horizontal vision reversal prisms. Ann. Otol. Rhinol. Lar. 90: 276–280 (1981).
7   Zee, D.S.; Yee, R.D.; Cogan, D.G.; Robinson, D.A.; Engel, W.K.: Ocular motor abnormalities in hereditary cerebellar ataxia. Brain 99: 207–243 (1976).

T. Yagi, MD, Department of Otolaryngology, Nippon Medical School,
1-1-5 Sendagi, Bunkyo-ku, Tokyo 113 (Japan)

Adv. Oto-Rhino-Laryng., vol. 30, pp. 13–16 (Karger, Basel 1983)

Influence of Cervical Reflexes on the Vestibulo-Ocular Reflex[1]

S. Vesterhauge, A. Månsson, K. Zilstorff

University ENT Department, Rigshospitalet, Copenhagen, Denmark

Introduction

One of the most common complaints in daily ENT practice is dizziness caused by head turning, usually ascribed to vertebrobasilar insufficiency. In order to construct a vestibular test procedure sensitive to this phenomenon, we have designed a test involving this type of motion. The procedure includes voluntary, active head rotations with a fixed positional amplitude of 60°, recording of horizontal EOG and comparison of the results of this procedure with the result of a similar, passive centrifuge stimulus. A similar method has been described by *Barnes* [1], different from ours concerning the stimulus amplitude and by our use of closed eyes.

Material and Methods

10 voluntary subjects without any history of vestibular disease, with normal caloric response and normal audiometry were tested. 0.2, 0.4 and 0.8 Hz rotations with a fixed amplitude of 60° were passively performed in a Servo Med CF 10 Centrifuge controlled by a PDP 8 Computer program. Active rotations were performed under guidance of a frequency modulated sound signal. The subjects were told to rotate their heads approximately 60°. Both tests were done with the eyes closed in a white cylindrical test box. The EOG signal was amplified and recorded simultaneously with a signal from a head-mounted accelerometer on an FM tape recorder. After computer elimination of the fast

[1] This study was supported by a grant from the Danish Space Board.

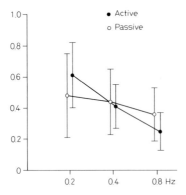

Fig. 1. The gain response of the active and passive stimuli. Mean ± SD.

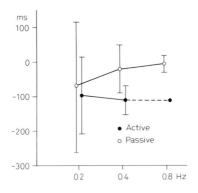

Fig. 2. The phase deviation of the oculomotor response to active and passive stimuli in the time domain. Mean ± SD. In the active stimulus at 0.8 Hz, only two recordings contained significant power at the stimulus frequency.

phases, reconstruction by extrapolation of the EOG signal and double integration of the accelerometer signal, a cross power spectral analysis was performed. The results appeared as gain and phase difference of the VOR compared with the stimulus at each frequency tested.

Results

Figure 1 demonstrates the gain response with the two different stimuli. No significant frequency dependence can be demonstrated with the passive stimulus. With the active stimulus, a correlation analysis shows a significant decrease in the gain with increasing frequency (r = –0.6, p < 0.01).

Figure 2 shows the phase response, the phase deviation being transformed into the time domain. It appears from the figure that the active procedure causes a time delay of the response at approximately 0.1 s, whereas no significant time delay appears with the passive stimulus. This delay seems to be independent of the frequency.

Discussion

In order to chose a realistic frequency range, all subjects were asked to shake their heads with a natural and not unpleasant frequency. The mean frequency was 0.59 Hz (SD 0.20). Because of troubles with the analysis of frequencies above 0.8 Hz, we decided to chose the range 0.2–0.8 Hz. *Barnes and Forbat* [2] used frequencies as high as 1.3 Hz. Another difference from their method is the fact that our subjects were told to close their eyes. This was decided under influence of the results of *Takemori and Suzuki* [3], who have reported that eye deviations due to neck torsion only appeared with eyes closed and not with eyes covered. If our results are compared with the results of *Barnes and Forbat* [2], no significant difference can be seen, neither in the gain nor in the phase in the passive stimulus. In the active stimulus, *Barnes* [1] reports a slightly higher gain, where we find a frequency dependence. It is a question whether this difference is due to the eye closure or to the higher positional amplitude in our study at higher frequencies (at 0.8 Hz: 60 vs. 12°).

Our study does not permit any conclusions about the central mechanisms of interaction between the cervico-ocular and the vestibulo-ocular reflexes. Seen from a technical point of view, it is tempting to compare the interaction with the effect of an analog Bessel filter. It has been suggested that the two reflex responses are algebraically summed during head movements [4]. If this is the case in our stimulus design, the gain effect of the neck afferents is frequency dependent. Somehow, a system description like our filter explanation to a higher extent fits the current view on brain stem function.

We are curious to know if we can carry on with this type of description when a group of patients with vertebrobasilar insufficiency has been examined.

Conclusions

In active head rotations, the gain decreases with increasing frequency. In passive rotations, the gain is more constant and independent of frequency. In active head rotations phase lag increases with increasing frequency – but has a constant time delay. In passive rotations, the time delay is almost zero. The effect of the neck afferents and the VOR can be compared with the effect of a low-pass Bessel filter.

References

1   Barnes, G.R.: Head-eye coordination in normals and in patients with vestibular disorders. Adv. Oto-Rhino-Laryng., vol. 25, pp. 197–201 (Karger, Basel 1979).
2   Barnes, G.R.; Forbat, L.N.: Cervical and vestibular afferent control of oculomotor response in man. Acta oto-lar. *88:* 79–87 (1979).
3   Takemori, S.; Suzuki, J.: Eye deviations from neck torsion in humans. Ann. Otol. *80:* 439–444 (1971).
4   Meiry, J.L.: The vestibular system and human dynamic space orientation (NASA CR-628, Washington 1966).

S. Vesterhauge, MD, ENT Department 2074, Rigshospitalet,
DK–2100 Copenhagen Ø (Denmark)

Adv. Oto-Rhino-Laryng., vol. 30, pp. 17–26 (Karger, Basel 1983)

Cutaneo-Pressure Reflex and Neck-Torsion Nystagmus in Rabbits

Manabi Hinoki, Shinsuke Ito

Department of Otolaryngology, Faculty of Medicine, Kyoto University, Kyoto, Japan

Introduction

Takagi [1953] reported that nystagmic responses due to a lesion of the unilateral labyrinth can be suppressed when rabbits are given pressure stimulation on the skin, particularly that of the proximal part of the auricle. With reference to the report of *Adrian and Zotterman* [1926], *Takagi* [1953] stated that tonic impulses from the pressure receptors of the skin probably inhibit development of a nystagmus, particularly its rapid phase. However, systematic investigations have not been directed to the role of the exteroceptors in the appearance of nystagmus of cervical origin or on what parts of the skin are responsible for alteration of the nystagmus of this type when cutaneo-pressure stimulation is given. The reason for the lack of such investigations is that there were not pertinent procedures for eliciting a nystagmic response of cervical origin. With our newly developed procedures we attempted to obtain a better understanding of these problems.

Experimental

Method for Induction of Nystagmus Due to Torsion of the Neck in Rabbits. A cloth is wrapped around the trunk and the rabbits are suspended in air with their four limbs not touching the ground. The head is then rotated clockwise around the longitudinal axis of the body at about 90°. The nystagmus thus induced was termed neck-torsion nystagmus.

The Electronystagmograph (ENG), Electromyograph (EMG) and Electroencephalograph (EEG) Measured Herein. Neck-torsion nystagmus was recorded on the right eye using the ENG and also in combination with either the EMGs or the EEGs. The EMGs were delivered from the deep group of the bilateral neck muscles. The EEGs were delivered from the neocortex and hippocampus.

Cutaneo-Pressure Stimulation. To avoid the possible concomitant of a reflex due to pain of the skin, cutaneo-pressure stimulation was given by wrapping a cuff of a pediatric sphygmomanometer around the proximal area of the auricle and inflating it up to a pressure of 200 mm Hg. The same procedure was applied to the skin of the hindlimbs.

Results

Effects of Cutaneo-Pressure Stimulation on Neck-Torsion Nystagmus and the EMGs from the Deep Neck Muscles

Results from Experiments with Cutaneo-Pressure Stimulation of the Auricle. Figure 1 shows neck-torsion nystagmus and EMG discharges from the deep neck muscles in a rabbit and with and/or without cutaneo-pressure stimulation of the auricle, on the lower side. Increase in EMG discharges from the right neck muscles due to torsion of the neck was inhibited when cutaneo-pressure stimulation was applied to the auricle, on the lower side. Thus, differences in the activity between the right and left neck muscles decreased, and such paralleled decreased neck-torsion nystagmus (fig. 1, upper part). After removal of the above-mentioned pressure stimulation, EMG discharges of the right neck muscles increased when the animal's neck was twisted clockwise. These significant differences in the activity between the right and left neck muscles led to a clearer appearance of neck-torsion nystagmus (fig. 1, lower part).

Figure 2 shows neck-torsion nystagmus and EMG discharges from the deep neck muscles in the above-mentioned rabbit and with and/or without cutaneo-pressure stimulation of the auricle, on the upper side. No significant inhibition of the EMGs was noted in the right neck muscles when the neck was twisted clockwise and with cutaneo-pressure stimulation of the auricle, on the upper side. Thus, there were significant differences in the activity between the right and left neck muscles and such paralleled the active appearance of neck-torsion nystagmus. After removal of the above-mentioned pressure stimulation, the results showed a similarity with regard to neck-torsion nystagmus and the EMGs. We obtained similar findings in 4 other rabbits.

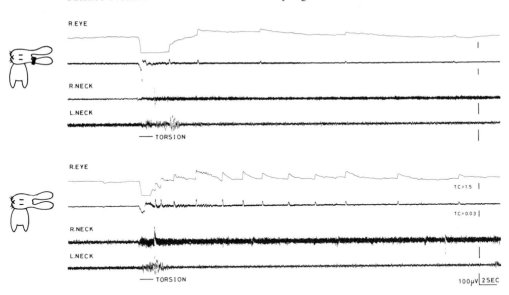

Fig. 1. Neck-torsion nystagmus and EMG discharges from the deep neck muscles in a rabbit and with and/or without cutaneo-pressure stimulation of the auricle, on the lower side. Increase in the EMGs of the right neck muscles due to torsion of the neck was inhibited when the animal was given cutaneo-pressure stimulation, on the lower side. Such a procedure led to a significant suppression of the nystagmus referred to herein. After removal of the above-mentioned pressure stimulation, increase in the EMGs of the same muscles due to torsion of the neck clearly appeared and paralleled the active appearance of the nystagmus of this type.

Results from Experiments with Cutaneo-Pressure Stimulation of the Hindlimbs. We found that effects of cutaneo-pressure stimulation of the hindlimbs on development of neck-torsion nystagmus can be readily evaluated when using rabbits with both decreased neck-torsion nystagmus and a sluggish arousal state in the EEGs. Therefore, the neck was twisted repeatedly with short intervals, of about 20 s. These so-prepared rabbits were tested in the following experiments.

Figure 3 shows neck-torsion nystagmus and the EMGs from the deep neck muscles in a rabbit given cutaneo-pressure stimulations on the hindlimbs. Little evidence of alteration was found with regard to nystagmus and the EMGs when the neck was twisted clockwise and with cutaneo-pressure stimulation of the hindlimb, on the side of the skull. Thus, no obvious neck-torsion nystagmus was observed in the ENG (fig. 3, upper part). Conversely, the activity of the right neck

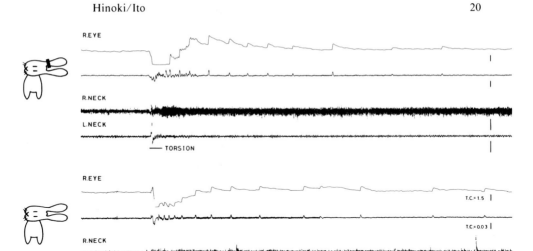

Fig. 2. Neck-torsion nystagmus and EMG discharges from the deep neck muscles in the same rabbits described in figure 1 and with and/or without cutaneo-pressure stimulation of the auricle, on the upper side. Increase in the EMGs of the right neck muscles was clearly induced under both conditions and paralleled the active appearance of the nystagmus referred to herein.

muscles due to torsion of the neck was reinforced when cutaneo-pressure stimulation was applied to the hindlimb, on the side of the chin. Thus, differences in the activity between the right and left neck muscles increased and led to the appearance of neck-torsion nystagmus (fig. 3, lower part). We obtained similar findings in 4 other rabbits.

Effects of Cutaneo-Pressure Stimulation on Neck-Torsion Nystagmus and the EEGs

Results from Experiments with Cutaneo-Pressure Stimulation on the Auricles. We found that the arousal state in the EEGs due to torsion of the neck decreased when cutaneo-pressure stimulation was applied to the auricle, on the lower side. In parallel with this, decreased neck-tor-

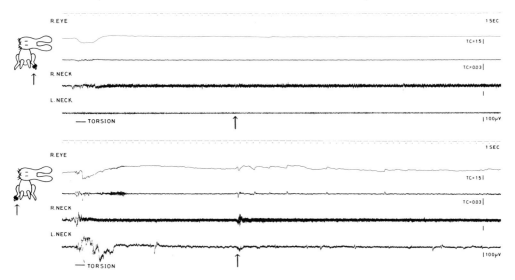

Fig. 3. Neck-torsion nystagmus and EMG discharges from the deep neck muscles in a rabbit and with cutaneo-pressure stimulation on the hindlimbs. Little evidence of alteration was induced with regard to neck-torsion nystagmus and EMGs when the animal was given cutaneo-pressure stimulation, on the side of the skull. Conversely, increase of the EMGs due to torsion of the neck was reinforced in the right neck muscles when the animals were given cutaneo-pressure stimulation, on the side of the chin. Such procedure led to the appearance of neck-torsion nystagmus.

sion nystagmus was induced. Conversely, little evidence of alteration was found with regard to the above-mentioned arousal state when cutaneo-pressure stimulation was applied to the auricle, on the upper side. Thus, neck-torsion nystagmus could be observed clearly (fig. 4). We confirmed these results in all the 5 animals examined.

Results from Experiments with Cutaneo-Pressure Stimulation on the Hindlimbs. We found that the arousal state in the EEGs was induced as a result of cutaneo-pressure stimulations applied to the hindlimbs on the sides of the skull and chin. This state was more marked in the case of latter stimulation than in the former. Furthermore, in the former stimulation, there was little evidence of nystagmic responses, whereas in the latter, neck-torsion nystagmus clearly appeared (fig. 5). We obtained similar findings in all the rabbits examined.

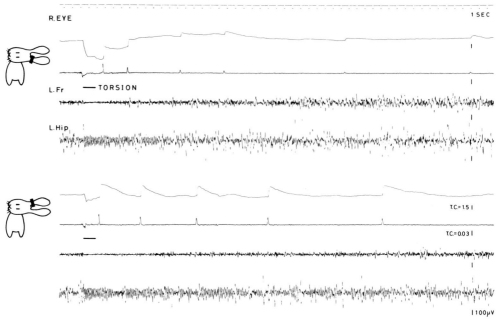

Fig. 4. Neck-torsion nystagmus and the EEGs in a rabbit and with cutaneo-pressure stimulation on the auricles. The arousal state in the EEGs due to torsion of the neck decreased when cutaneo-pressure stimulation was applied to the auricle, on the lower side. Conversely, little evidence of alteration was found with regard to the above-mentioned arousal state when the same stimulation was applied to the auricle, on the upper side. Thus, in the former stimulation, decreased neck-torsion nystagmus was induced, whereas in the latter, such a nystagmus clearly appeared.

Comment

Takagi [1953] reported that nystagmic responses due to a lesion of the unilateral labyrinth are suppressed when animals are given cutaneo-pressure stimulation. Conversely, tactile stimulation induces activation of this type of nystagmus. Citing the report of *Adrian and Zotterman* [1926], *Takagi* [1953] explained that tactile stimulation favorably acts on rapidly adapting components involved in the skin receptors, through which activation of nystagmus is induced. In contrast, pressure stimulation affects slowly adapting components of the skin receptors, through which decreased nystagmus is brought about. Since we found that supporting the hindlimbs with a bar and/or a plate there was an increase in the neck-torsion nystagmus as well as an increased

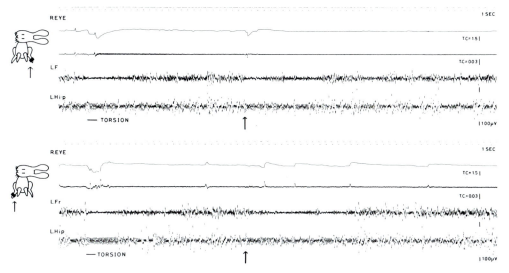

Fig. 5. Neck-torsion nystagmus and the EEGs in a rabbit and with cutaneo-pressure stimulation on the hindlimbs. The arousal state in the EEGs was induced as a result of cutaneo-pressure stimulations applied to the hindlimbs on the sides of the skull and chin. This state was more marked in the case of latter stimulation than in the former. Furthermore, in the former stimulation, there was little evidence of nystagmic responses, whereas in the latter, neck-torsion nystagmus clearly appeared. The arrows in this figure indicate the beginning of cutaneo-pressure stimulation on the hindlimbs.

activity of the deep neck muscles and a clearer arousal state in the EEGs, we wondered if the cutaneo-pressure stimulation would constantly inhibit the nystagmic responses.

In this condition, the pressure receptors of the skin in the hindlimbs are to be stimulated, together with the proprioceptors of these limbs. Thus, we speculated that effects of the cutaneo-pressure reflex possibly vary with regard to development of neck-torsion nystagmus, depending on the sites of the stimulation given. In the present investigation, we found that the cutaneo-pressure reflex elicited in the auricle reduced the increased activity of the deep neck muscles contralaterally, and that such states paralleled decreased neck-torsion nystagmus and a sluggish arousal state in the EEGs. In contrast, the cutaneo-pressure reflex elicited in the hindlimbs reinforced the increased activity of the neck muscles ipsilaterally and such states led to activation of neck-torsion nystagmus and a clearer arousal state in the EEGs. Recently, we also found that cutaneo-pressure stimulation elicited in the trunk induced decreases of neck-torsion nystagmus and moreover, such effects were more marked in the case of stimulation of the trunk on the side of the chin than on that of the skull. Furthermore, little evidence of effects was found with regard to development of neck-torsion nystagmus when animals were given cutaneo-pressure stimulation on the forelimbs. It is worth noting that the cutaneo-pressure reflex plays a certain role in enhancement and/or suppression of neck-torsion nystagmus, in accordance with the sites of the stimulation given. We now come to the question of what neural elements are relevant to suppression and/or activation of neck-torsion nystagmus, in the case of cutaneo-pressure stimulation.

Before explanation of the above-mentioned problem, it seems necessary to mention the neural mechanism of neck-torsion nystagmus. *Hikosaka and Maeda* [1973] reported that deviation of the eyes due to torsion of the neck is induced on the basis of the following neural mechanism. That is, afferent volleys from the neck joints ascend ipsilaterally in the spinal cord, cross to the contralateral side in the brain stem and project to the vestibular nuclei, thus interacting with the vestibulo-abducens reflex activity.

Richmond and Abrahams [1979] reported that in cats, joint receptors are absent in the upper cervical joints, and that abundant muscle spindles are distributed in the neck muscles near the joints referred to herein. Thus, afferent volleys from the deep neck muscles probably participate in the production of eye movement of this type. However, detailed explanations with regard to the neural mechanism of the rapid phase of neck-torsion nystagmus were not given. From known fiber connections in the central nervous system, the spinoreticular tract seems important in eliciting rapid phase of nystagmus of this type, since fibers of this tract ascend along the lateral fascicle and the anterior column and terminate in the reticular formation both the medulla oblongata and the pons [*Brodal*, 1957] and moreover, some fibers of this tract ascend directly to the midbrain, connecting to the Deiter's nucleus [*Terada*, 1960]. Furthermore, the reticular formation of the brain stem, particularly the paramedian zone of the pontine reticular formation (PPRF) reportedly plays a major role in the formation of the rapid phase of nystagmus [*Teng* et al., 1958]. From these reports, it is probable that the above-mentioned phenomena regarding neck-torsion nystagmus develop in connection with the following two groups of neural elements and such elements are arranged in the following way: (1) The pressure receptors of the hindlimbs → posterior roots → ipsilateral long ascending fibers in the posterior funiculus → synaptic connection with motoneurons in the cervical cord → the deep neck muscles → the proprioceptors distributed in the deep neck muscles → the spinoreticular tract → the vestibular nuclei and reticular formation of the brain stem. The pathways referred to herein are mainly relevant to excitatory effects of the cutaneo-pressure reflex on development of neck-torsion nystagmus, since activation of neck-torsion nystagmus developed in combination with ipsilaterally increased EMG discharges of the deep neck muscles and the arousal state in the EEGs and such was induced when rabbits were given cutaneo-pressure stimulation of the hindlimb, on the side of the chin. (2) The pressure receptors of the auricle → posterior roots → crossed association neurons → synaptic connections with motoneurons in the cervical cord → the deep neck muscles → the proprioceptors distributed in the deep neck muscles → the spinoreticular tract → the vestibular nuclei and reticular formation of the brain stem. The pathways referred to herein are mainly relevant to inhibitory effects of the cutaneo-pressure reflex on development of neck-torsion nystagmus, since inhibition of neck-torsion nystagmus appeared in combination with contralaterally reduced EMG discharges of the neck muscles and a sluggish arousal state in the EEGs and such was induced when animals were given cutaneo-pressure stimulation of the auricle, on the lower side.

Further investigations are under way to substantiate the above-mentioned postulation and which would account for the activation and/or suppression of neck-torsion nystagmus in the case of cutaneo-pressure stimulation.

Development of the nystagmus referred to herein probably relates with both the neck proprioceptors and the otolith. Therefore, the neural mechanism underlying activation and/or suppression of neck-torsion nystagmus possibly includes effects of the cutaneo-pressure reflex not only on the neck proprioceptors, but also on the otolith.

We found that when the otolith only was stimulated, nystagmic responses were never induced. In contrast, nystagmic responses were induced when the trunk was twisted in a natural head position. Thus, we assume that the alteration of neck-torsion nystagmus referred to herein is mainly due to increased and/or decreased activities of the neck proprioceptors, which are brought about by the above-mentioned cutaneo-pressure stimulation.

Conclusions

From the above-mentioned results we conclude: (1) Cutaneo-pressure reflex plays a certain role in enhancement and/or suppression of neck-torsion nystagmus. (2) The cutaneo-pressure reflex elicited in the auricle reduces the increased activity of the deep neck muscles contralaterally, whereas that elicited in the hindlimbs reinforces the increased activity of the same muscles ipsilaterally. (3) The former cutaneo-pressure reflex induces decreased activity of the brain through decreases in the activity of the neck muscles. In contrast, the latter cutaneo-pressure reflex induces increased activity of the brain through increases in the activity of these same muscles. Thus, the former leads to decreased neck-torsion nystagmus, while the latter is related to increased neck-torsion nystagmus.

Acknowledgements

We are greatly indebted to Prof. *K. Niimi* (Department of Neuro-Anatomy, Okayama University) for valuable advice regarding fiber connections in the central nervous system, and to *M. Ohara*, Kyushu University, for editorial assistance.

References

Adrian, E.D.; Zotterman, Y.: The impulses produced by sensory nerve endings. Part 3. Impulses set up touch and pressure. J. Physiol. *61:* 465–483 (1926).

Brodal, A.: The reticular formation of the brain stem, pp. 1–87 (Oliver & Boyd, Edinburgh, 1957).

Hikosaka, O.; Maeda, M.: Cervical effects on abducens motoneurons and their interaction with the vestibulo-ocular reflex. Exp. Brain Res. *18:* 512–530 (1973).

Richmond, F.J.; Abrahams, V.C.: What are the proprioceptors of the neck? Prog. Brain Res. *50:* 245–254 (1979).

Takagi, K.: Cutaneo-pressure reflex and labyrinthine function. Brain Res., Osaka *6:* 179–191 (1953).

Teng, P.; Shanzer, S.; Bender, M.B.: Effects of brain stem lesions on optokinetic nystagmus in the monkey. Neurology, Minneap. *8:* 22–26 (1958).

Terada, S.: Course and termination of the long ascending spinal tracts and of the medial leminiscus in cats, with special reference to the relation of these tracts to the reticular formation. J. Chiba med. Soc. *36:* 1065–1075 (1960).

M. Hinoki, MD, Department of Otolaryngology, Faculty of Medicine,
Kyoto University, Sakyo-ku, Kyoto 606 (Japan)

Adv. Oto-Rhino-Laryng., vol. 30, pp. 27–29 (Karger, Basel 1983)

Eye Position-Related Activity in Deep Neck Muscles of the Alert Cat

Pierre Paul Vidal[a], André Roucoux[b], Alain Berthoz[a], Marc Crommelinck[b]

[a] Laboratoire de Physiologie Neurosensorielle du CNRS, Paris, France;
[b] Laboratoire de Neurophysiologie de l'Université de Louvain, Bruxelles, Belgique

The activity of some neck muscles has been shown to be related to eye movements in different species [*Bizzi* et al., 1971; *Fuller*, 1981; *Roucoux* et al., 1981; *Vidal* et al., 1982]. These muscles have also been shown to receive specific vestibular projections [*Peterson* et al., 1981].

The aim of the present work was to analyze the activity of these neck muscles, especially the small and deep ones, in relation with eye movements. We shall concentrate here on the description of the results obtained on head-fixed cats, for horizontal eye deviations.

The methods have been described in detail in *Vidal* et al., [1982]. Eye movements are recorded by the coil technique in alert, intact cats, whose head and body are firmly restrained. Electromyograms (EMG) are recorded in neck muscles by chronically implanted bipolar electrodes. Tape-recorded EMG is rectified and integrated (5–50 ms time constant). Eye movements are obtained either spontaneously in the light or darkness, evoked in complete darkness by horizontal rotation of the animal or by rear projection on a screen of a moving random-dot pattern.

The activity of all the investigated muscles is modulated by horizontal eye position. For example, a right neck muscle increases its frequency discharge as the eye adopts increasing eccentricities towards the right, beyond a certain threshold. The left muscle symmetrically decreases its activity.

During vestibular or optokinetic nystagmus, the same relationship is observed. Figure 1 shows an example of recordings obtained during vestibular stimulation in the horizontal plane. The EMG has been inte-

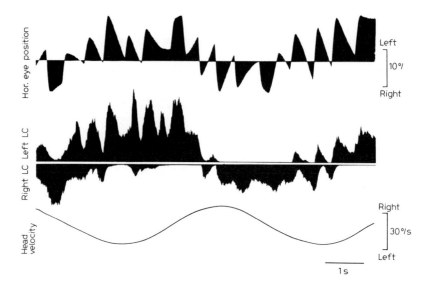

Fig. 1. Integrated EMG of both longissimus capitis muscles (LC) during sinusoidal horizontal head rotation in darkness. Right muscle EMG is shown upside down for better visual comparison with eye position.

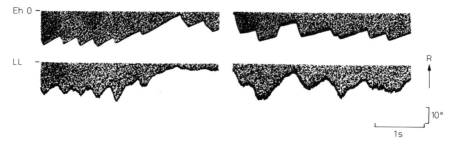

Fig. 2. Comparison between horizontal eye position during rightward optokinetic stimulation and rectified, integrated EMG of left longissimus capitis muscle (ii).

grated with a 50 ms time constant. The activity of right and left longissimus can be compared with the horizontal eye position trace and the head velocity signal. The line drawn through the eye position trace indicates midposition in the orbit. Clearly, muscle activity appears to be related to eye position and not to the vestibular signal.

Figure 2 illustrates the rectified EMG of the left longissimus capitis muscle during optokinetic stimulation to the right. The upper limit of the shaded areas corresponds to the eye midposition and to the integrated EMG baseline. EMG is inverted for better visual comparison. Note the close correlation between the muscle activity and eye position.

Most of the neck muscles studied so far exhibited similar behaviors. The best relationships between eye position and muscle activity were obtained from the longissimus capitis, obliquus capitis cranialis and rectus capitis dorsalis and, to a lesser extent, splenius, complexus and biventer cervicis. Muscle activity related to vertical eye position will not be described here.

In our experimental situation, the vestibulocollic reflex signal as described by *Peterson* et al., [1981] in the precollicular sectioned cat does not generally appear.

References

Bizzi, E.; Kalil, R.E.; Tagliasco, V.: Eye-head coordination in monkeys: evidence for centrally patterned organization. Science *173:* 452–454 (1971).

Fuller, J.H.: Eye and head movements during vestibular stimulation in the alert rabbit. Brain Res. *205:* 363–381 (1981).

Peterson, B.W.; Bilotto, G.; Goldberg, J.; Wilson, V.J.: Dynamics of vestibulo-ocular, vestibulo-collic and cervico–collic reflexes. Ann. N.Y. Acad. Sci. *374:* 395–402 (1981).

Roucoux, A.; Crommelinck, M.; Guitton, D.: The role of superior colliculus in the generation of gaze shift; In Fuchs, Becker, Progress in oculomotor research, developments in neuroscience, vol. 12, pp. 129–135 (Elsevier/North-Holland Biomedical Press Amsterdam 1981).

Vidal, P.P.; Roucoux, A.; Berthoz, A.: Horizontal eye position-related activity in neck muscles of the alert cat. Exp. Brain Res. *46:* 448–453 (1982).

A. Roucoux, MD, Université Catholique de Louvain, Faculté de Médecine, Laboratoire de Neurophysiologie, Avenue Hippocrate, 54, UCL/5449, B-1200 Bruxelles (Belgium)

Adv. Oto-Rhino-Laryng., vol. 30, pp. 30–33 (Karger, Basel 1983)

Somatosensory Nystagmus: Physiological and Clinical Aspects

Willem Bles[a], Theo Klören[a], Wolfgang Büchele[b], Thomas Brandt[b]

[a] ENT Department, Free University Hospital, Amsterdam, The Netherlands;
[b] Neurological Clinic with Clinical Neurophysiology, Alfried Krupp Hospital, Essen, FRG

Introduction

Somatosensory signals from joint as well as from musculotendinuous receptors provide an accurate kinaesthetic feedback of the extent of limb movements. They contribute to the perception of self-motion during active locomotion and converge with vestibular and visual afferences upon multimodal neurons in the vestibular nuclei and thalamus which project to cortical sensory areas in the anterior parietal lobe. It could be experimentally demonstrated that in man movement of the limbs may induce a compelling illusion of self-motion and a purely somatosensory nystagmus, even in the absence of concurrent vestibular or optokinetic stimulation: (1) arthrokinetic nystagmus (AKN) is evoked in objectively stationary subjects seated in darkness inside a rotating cylinder, when they are tracking the rotation of the cylinder by placing their hands on its inner wall [*Brandt* et al., 1977]; (2) apparent stepping around nystagmus (ASAN) evoked in stationary subjects on a small circular treadmill in darkness [*Bles*, 1981].

Somatosensory Nystagmus in Normals

AKN and ASAN have a latency of up to several seconds after stimulus onset and show a build up of slow phase velocity (SPV). The mean eye position ('Schlagfeld') deviates in the direction of the fast phase and following the end of stimulation positive after-nystagmus can be

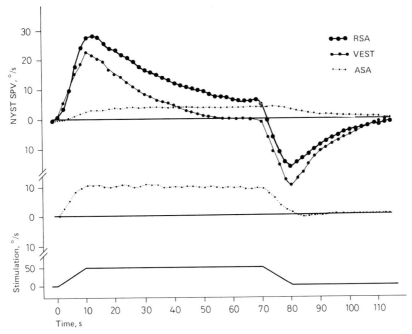

Fig. 1. Mean nystagmus slow phase velocity from both CW and CCW rotation during real stepping around (RSA), which is a combination of vestibular and somatosensory stimulation, during pure vestibular stimulation (VEST) and during pure somatosensory stimulation, i.e. apparent stepping around (ASA), obtained from 16 healthy controls (upper part) and 7 patients with bilateral loss of vestibular function (lower part). Stimulus pattern is shown in the lower part.

seen, similar to positive optokinetic after-nystagmus. The mean maximum SPV (n = 18) at stimulus speeds of 10, 30 and 60°/s was 3.6, 4.3 and 5.3°/s for AKN and 7.2, 10.6 and 10.5°/s for ASAN. For both stimulus conditions the gain decreases with increasing stimulus speed (AKN: 0.4–0.1; ASAN: 0.7–0.2). The increased nystagmus SPV with real stepping around (RSA) (simultaneous somatosensory and vestibular stimulation) as compared to passive body rotation (pure vestibular stimulation) can be explained by the contribution of somatosensory signals as shown by the ASAN in the upper part of figure 1. However, SPV does not result from purely algebraic summation, suggesting facilitation in case of congruent sensorial information.

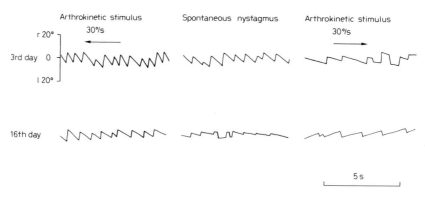

Fig. 2. Course of spontaneous nystagmus (middle) and arthrokinetic nystagmus in a female patient (U.K., age 20 years) with an acute lesion of the left labyrinth 3 and 16 days after onset of symptoms. Note that after 3 days an arthrokinetic stimulus of 30°/s to the right does not overrule the spontaneous nystagmus to the right. At 16 days, arthrokinetic stimulation to the right induces nystagmus to the left, which overrules the sequelae of the partially compensated labyrinthine defect.

Somatosensory Nystagmus in Patients

Patients with labyrinthine lesions exhibit characteristic abnormalities of somatosensory nystagmus.

Bilateral Loss of Labyrinthine Function (fig. 1, lower part). (1) Drastic shortening of latencies for ASAN (and circular vection); (2) rapid build up of ASAN slow phase velocity; (3) increased gain of ASAN, and (4) absence of ASA after-nystagmus (cf. the earlier reported absence of optokinetic after-nystagmus in these patients).

These alterations may be interpreted as a compensation for the loss of vestibular function: an increase of the particular sensorial weight attributed to the somatosensory information on self-motion cues. In these patients the rapid onset and rapid build up of the somatosensory nystagmus may result from the absence of vestibular signals, which, if present, generate a signal that conflicts with the somatosensory response.

Unilateral Labyrinthine Lesions (fig. 2). (1) Spontaneous nystagmus biases the somatosensory nystagmus; (2) the modulation is asymmetric, being stronger in the direction of the spontaneous nystagmus (about 15%, calculated for 13 patients with ASAN), and (3) the gain of the somatosensory nystagmus is increased. These findings also suggest a change of the sensorial weight given to the somatosensory information similar to the patients with bilateral loss of vestibular function.

Differential Effects of Hemispheric Lesions on Optokinetic and Somatosensory Nystagmus. We have the impression that the supratentorial visual system is not important in the generation of the somatosensory nystagmus since hemispheric lesions (n = 4) with directional preponderance of OKN (> 30%) did not affect somatosensory nystagmus.

References

Bles, W.: Stepping around: Circular vection and coriolis effects; in Long, Baddeley, Attention and performance, vol. IX, pp. 47–61 (Lawrence Erlbaum Associates, Hillsdale 1981).

Brandt, T.; Büchele, W.; Arnold, F.: Arthrokinetic nystagmus and ego-motion sensation. Exp. Brain Res. *30:* 331–338 (1977).

Dr. W. Bles, ENT Department, Free University Hospital,
de Boelelaan 1117, NL–1007 MB Amsterdam (The Netherlands)

Adv. Oto-Rhino-Laryng., vol. 30, pp. 34–39 (Karger, Basel 1983)

A Mechanical View of
Peripheral Vestibular Function

Juergen Tonndorf[1]

Fowler Mem. Laboratory, Department of Otolaryngology, College of Physicians and Surgeons of Columbia University, New York, N.Y., USA

This paper will examine, or re-examine respectively, some aspects of the mechanical stimulation and responses of the vestibular end organs: (1) the different effects of single otoliths and multiple otoconia; (2) the relevant input quantity determining end-organ responses, and (3) the cause of the long duration of postrotatory nystagmus. Central, neural processing will not be considered, since that might unnecessarily complicate the issues at hand. It may be added, however, that there is some recent evidence, at least for the cochlear organ, indicating that efferent nerve action might directly modify peripheral mechanical responses

Macular Stimulation

Figure 1 shows, in a schematic manner, the mode of hair cell stimulation caused by a shift in position of a large otolith as found in fishes. Because of the strong friction in the capillary space underneath, appreciable shearing forces are brought about only around the margins of the otolith, i.e., it is predominantly here that hair cells are being stimulated. This situation indicates the benefits accruing when the solitary, large otolith of fishes is subdivided into a larger number of smaller otoconia as present in higher vertebrates: The number of marginal regions in the sense of figure 1 increases and, therefore, more hair cells are being stimulated; consequently, the sensitivity of the organ as a whole ought to improve. Furthermore, an increase in the number of responding receptors, especially if they have slightly different thresholds, ren-

[1] Supported by a grant from The Guinta Foundation, Hackensack, N.J., USA.

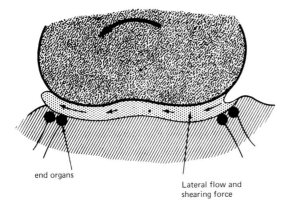

end organs

Lateral flow and
shearing force

Fig. 1. For explanations see text.

ders the input/output function of the entire organ more gradual as was demonstrated for assemblies of tactile receptors. Hence, the differential sensitivity should also be improved.

Cupular Reception of Energy
It has been argued that the cupula is a pressure receiver. While this is quite correct under certain restrictions, as we will see, it does not cover the general case, which will be presented: Angular acceleration of a semicircular canal produces a pressure action on the cupula, on account of the moment of inertia of the endolymphatic fluid. (An actual flow is prevented by the cupular seal on the ampullar roof.) This pressure causes a displacement in proportion to the elastic restoring force of the cupula. Clearly, some pressure must be exerted before a displacement may ensue. The product of pressure and displacement is proportional to the energy received. Energy is needed to do work, i.e., to generate receptor potentials inside the hair cells. Displacements will be small if impedance is high and vice versa. One may therefore also consider the work done against a certain impedance (Z), in which case work (W) becomes

$$W = \frac{p^2}{Z} T,$$

where p = pressure, T = the time period, during which the pressure is acting, A = the area, over which it is applied. For a specific input impedance, therefore, the energy received can be completely described in terms of pressure. Moreover, since the input impedance is rather high (as is that of virtually all signal receivers, man-made as well as biological ones) displacement is quite small and can be neglected in first approximation; thus, at least in the intact organ, pressure is an adequate descriptor of cupular and/or hair-cell responses. However, mere displacement data by themselves are not meaningful, since displacement (the mechanical equivalent of electrical charge) is incapable of doing any work, even the minute amounts required in the present case. The problem for the experimenter is that work, or energy, and pressure can, at present, be assessed only for entire organs, and even then with difficulty. For individual receptors, however, displacement is the only quantity that can be measured at the current state of the art.

Duration of Postrotatory Nystagmus

Postrotatory nystagmus in a typical Bárány chair test, on the average, lasts approximately 30 s. The fact that the ampullar nerve is known to continue discharging as long as the cupula and the attached hair-cell stereocilia stay deflected was taken to suggest that after its deflection the cupula should very slowly return to its resting position. In terms of a torsion-pendulum model, the action of which is completely determined by the restoring force of the cupula, the moment of inertia of the endolymph and its damping, this left but one conclusion, i.e., that the restoring force ought to be rather weak and damping very strong and quite likely overcritical.

Intuitively, this conclusion is not wholly convincing. If damping were really that high, semicircular canals should be very sluggish in responding to any stimulation, since the time constant of the on-period ought to be about as long as that of the off-period. Brief, transient stimuli might have long ceased when the system would at last be capable of responding, if indeed it could do so at all. Under such conditions, compensating motor reflexes would come much too late. Data on reflex latency argue against such an explanation. The latency of sceletal-muscle reflexes after vestibular stimulation is given as 12–14 ms and that of postrotatory nystagmus in the horizontal plane as immeasurably short on strong stimulation, although extending up to 2–3 s on weak stimulation; but even the latter values are much shorter than the duration of the postrotatory event would suggest.

We have to take a look at the results of tests on horizontal torsion chairs. Figure 2 presents frequency response curves obtained with sinu-

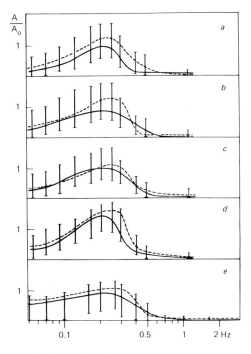

Fig. 2. For explanations see text.

soidal inputs over a frequency range from 0.05 to 2 Hz on two vestibularly normal subjects in a variety of testing situations. In all instances
there was a definite response maximum, i.e., a resonant point, most frequently at about 0.3 Hz, indicating that damping was less than critical.
(For critical damping, the curve would have sloped down monotonically from the low frequencies on, reaching a value of one-half at the
resonant point.) The damping factor d of the curves shown in figure 2
is estimated to be approximately 3.14.

Phase assessments gave essentially the same results. Figure 3 presents phase curves from two independent studies employing torsion-
chair testing. The first one (n = 25) used sinusoidal accelerations at five
frequencies between 0.01 and 0.16 Hz, the second one (n = 30) pseudorandom accelerations and evaluation for five frequencies between
0.02 and 1 Hz. Both curves are plotted onto a grid of normalized phase
curves of a hypothetical resonant system, calculated for various
amounts of damping. Their resonant points were taken at 0.32 Hz. The

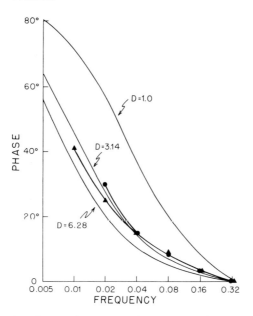

Fig. 3. ▲ = sine wave testing; ● = pseudorandom testing. For other information see text.

two curves run close to the phase curve calculated for d = 3.14. The discrepancy between them at frequencies below 0.04 Hz may have to do with the fact that at discrete frequency testing some habituation might have occurred when the half-periods became very long at lower frequencies, a shortcoming that is apparently not shared by pseudorandom testing.

A damping factor of about 3.14, as suggested by both the response curves of figure 2 and the phase curves of figure 3, makes the semicircular canals rather ideal, broad-band receptors, having a wide frequency response curve below a relatively small resonant point and rather short on-effects.

These results are clearly incompatible with the assumption that the long duration of postrotatory nystagmus would be determined by very strong, i.e., overcritical damping, as was suggested by the straightforward application of the torsion-pendulum model. The question then arises as to what may cause the pronounced difference between the results obtained by impulse testing in a Bárány chair and by sine wave (or pseudorandom) testing in a torsion chair?

For want of a suitable mechanical explanation, central processes have been invoked to account for the above discrepancies. Although modifications of nystagmus durations caused by central vestibular effects have been demonstrated, this author still sees a possibility that the long duration of postrotatory nystagmus might have primarily peripheral, mechanical causes. On low-magnitude stimulation, as in a torsion-chair test, the semicircular canals are expected to respond in a linear manner, but the sudden stop of a Bárány chair might constitute a severe overstimulation, resulting in a nonlinear response. That is to say, the elastic limits of the cupula might have been exceeded, producing a hysteresis effect that requires time to repair itself. There is good support for such hypothesis. Biological tissues with apparent elastic properties are in reality visco-elastic in nature displaying their divergent properties under different conditions. In the inner ear this has first been demonstrated for the tectorial membrane. Normally, this membrane is highly elastic. However, if perilymph, with its high Na content, was allowed to enter the cochlear duct, the membrane lost its elasticity and became viscous instead. This condition, called viscous deformation, is apparently reversible when the original K/Na equilibrium is restored. Such (reversible) changes can also be observed after overstretching, as was shown, for example, on the stereocilia of cochlear hair cells after sound exposure at levels producing temporary threshold shifts.

If the cupula undergoes viscous deformation under the effect of the strong impulse given by the sudden stopping of a Bárány chair, it should remain deflected until it has regained its restoring force, which apparently takes about 30 s. During this time, the concomitant deflection of the attached stereocilia should continue to generate hair-cell receptor potentials and, ultimately, action potentials in the ampullary nerve so that nystagmus is displayed. The period of nystagmus reversal that follows the primary postrotatory nystagmus appears to represent after-images of central origin and need not concern us here.

J. Tonndorf, MD, College of Physicians and Surgeons of Columbia University, Department of Otolaryngology, New York, NY 10032 (USA)

Adv. Oto-Rhino-Laryng., vol. 30, pp. 40–45 (Karger, Basel 1983)

Mathematical Model of the Slow Phase Caloric Nystagmus[1]

Zaid Chalabi

Royal National Throat, Nose and Ear Hospital and Engineering in Medicine Laboratory, Imperial College, London, England

Introduction

A number of mathematical models have attempted to describe vestibular dynamics in caloric stimulation [1, 2, 4, 6, 7]. A more rigorous analysis of the heat transport dynamics is now presented. It is assumed that the major heat transfer pathway is through the temporal bone [7] rather than the middle ear space [5].

The thermal stimulus induced in the external acoustic meatus causes heat to diffuse through the temporal bone to the semicircular canal (fig. 1) developing a torque across the canal axis due to temperature-induced variation of endolymph density in a gravitational field. Consequently, the cupula is deflected, and the discharge rate of the sensory cells is modulated eliciting, through different neural pathways, eye movements.

Model Description

Heat Flow in Temporal Bone

The temporal bone is a non-homogeneous and anisotropic medium. It is an air cell bone mixture perfused with blood with the petrous part of the temporal bone containing the semicircular canals. The temporal bone is viewed as a stationary rigid porous medium through

[1] This work was supported by the British Medical Research Council.

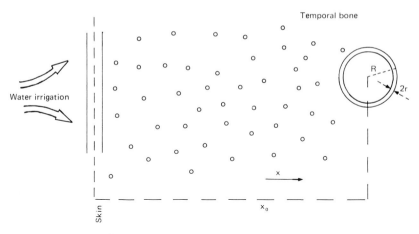

Fig. 1. Head diffusion from the external acoustic meatus to the semicircular canal.

which an incompressible fluid (blood) is flowing. The bone and blood are then viewed as a continuum and a single equation is used to describe the differential energy balance. The medium is assumed to have constant thermophysical properties averaged over the region.

A local volume-averaged differential energy balance equation is derived by averaging the energy differential equations in the solid and liquid phases and combining them into one single equation. Assuming a non-oriented porous medium and isotropic blood perfusion, and in the absence of blood accumulation or depletion, the bio-heat equation becomes [3]:

$$[F \cdot P_{BL} \cdot C_{BL} + (1-F) \cdot P_{BO} \cdot C_{BO}] \frac{\partial T}{\partial t} + F \cdot P_{BL} \cdot C_{BL} \cdot <\vec{W}> \cdot \text{grad } T =$$

$$[F \cdot K_{BL} + (1-F) \cdot K_{BO}] \nabla^2 T \tag{1}$$

Consider for simplicity a one-dimensional heat flow. Applying Laplace transformation to the reduced energy transport equation and to the appropriate boundary condition, gives the transfer function of the spatial distribution of the temperature across the temporal bone:

$$\frac{T(x,s)}{T_{i(s)}} = \frac{\sigma}{\sigma + \sqrt{\left(\frac{a}{b}\right)s + \left(\frac{c}{2b}\right)^2} - \frac{c}{2b}} \cdot \exp\left[-x\left(\sqrt{\left(\frac{a}{b}\right)s + \left(\frac{c}{2b}\right)^2} - \frac{c}{2b}\right)\right] \tag{2}$$

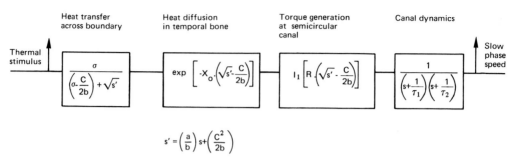

$$s' = \left(\frac{a}{b}\right)s + \left(\frac{c^2}{2b}\right)$$

Fig. 2. Transfer function of the slow phase caloric nystagmus. a, b and c are constants related to the thermophysical properties of the temporal bone.

Convective Endolymph Movement and Cupula Deflection

The semicircular canal is idealized as a toroid. The induced torque transfer function is given by [3]:

$$\frac{\Gamma(s)}{T_{i(s)}} = \alpha \cdot \frac{\sigma}{\sigma + \sqrt{\left(\frac{a}{b}\right)s + \left(\frac{c}{2b}\right)^2} - \frac{c}{2b}} \cdot \exp\left[-x_o\left(\sqrt{\left(\frac{a}{b}\right)s + \left(\frac{c}{2b}\right)^2} - \frac{c}{2b}\right)\right] \cdot$$

$$(I_1)\left[R \cdot \left(\sqrt{\left(\frac{a}{b}\right)s + \left(\frac{c}{2b}\right)^2} - \frac{c}{2b}\right)\right] \tag{3}$$

The torsion pendulum model of the canal dynamics [6] can be used to compute the transfer function of the cupular angular deflection:

$$\frac{CUP(s)}{\Gamma(s)} = \frac{\gamma}{\left(s + \frac{1}{\tau_1}\right)\left(s + \frac{1}{\tau_2}\right)} \tag{4}$$

Slow Phase Speed

Since the slow phase speed is proportional to cupular angular deviation [7], the overall transfer function of the caloric nystagmus is (fig. 2):

$$\frac{SPV(s)}{T_{i(s)}} = K \cdot \frac{CUP(s)}{\Gamma(s)} \cdot \frac{\Gamma(s)}{T_{i(s)}} \tag{5}$$

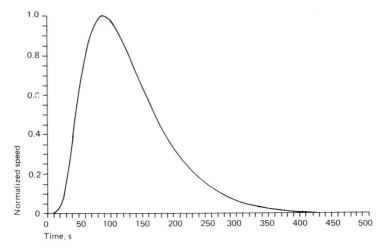

Fig. 3. Slow phase speed of caloric nystagmus.

Notation

Grad = Gradient; ∇^2 = Laplacian; I_1 = modified first order Bessel function; K_{BO} (K_{BL}) = thermal conductivity of bone (blood); C_{BO} (C_{BL}) = heat capacity of bone (blood); σ = heat transfer coefficient; T = temperature; τ_1, τ_2 = time constants related to the semicircular canal dynamics; γ = gain constant of the canal system; F = ratio of the volume occupied by blood in the averaging volume to the total volume; $<\vec{W}>$ = average blood velocity vector; a = $F \cdot C_{BL} \cdot P_{BL} + (1\text{-}F) \cdot C_{BO} \cdot P_{BO}$; b = $F \cdot K_{BL} + (1\text{-}F) \cdot K_{BO}$; c = $F \cdot C_{BL} \cdot P_{BL} \cdot <W>_X$; $P_{BO}(P_{BL})$ = density of bone (blood); T_i = input thermal stimulus; R = radius of the semicircular canal.

Model Simulation

Mean values of the physical constants are primarily taken from *Baertschi* et al. [1] and *Demers and Outerbridge* [4]. The slow phase responses in a group (each characterized by particular set of physical constants with one of the parameters as a variable) are referenced to one response in the group. The time course of the slow phase speed is obtained by convoluting the input stimulus (assumed to be an exponential decay rather than a step, to simulate real physical conditions of irrigation) with the impulse response derived from the caloric nystagmus transfer function (fig. 3). It is shown that better temporal bone thermal conductivity tends to increase the magnitude of the maximum

Table 1. Effect of thermophysical properties of the temporal bone on maximum slow phase speed (MSVS) and its latency (T_{MAX})

a	K_{BO}, $s^{-1}/°·cm$	MSVS	T_{MAX}, s	c	X_0, cm	MSVS	T_{MAX}, s
	0.0003	1.00	110		0.55	1.00	75
	0.0004	1.15	93		0.65	0.77	88
	0.0005	1.26	82		0.75	0.61	103
	0.0006	1.36	74		0.85	0.48	118

b	σ, cm^{-1}	MSVS	T_{MAX}, s	d	‹W›, cm^{-2}/s	MSVS	T_{MAX}, s
	2	1.00	110		0.0	1.00	89
	7	1.80	95		0.0001	0.93	90
	12	2.06	91		0.0004	0.75	91
	17	2.18	88		0.0008	0.54	93

slow phase speed and shortens its latency (table Ia) as does a large heat transfer coefficient between the irrigating medium and the bone (table Ib). The longer the transmission pathway from external acoustic meatus to semicircular canal, the weaker is maximum slow phase speed and longer is its latency (table Ic). The effect of blood circulation in the bone on the slow phase nystagmus is rather complex. Blood, being in motion, may convect heat in any direction with respect to the local temperature gradient. The driving force of heat transfer between blood and tissues and blood is mainly the difference between blood and tissue temperatures. It is shown that increased blood flow decreases maximum slow phase speed in thermal stimulation (table Id).

Conclusion

The significance of this study is to show the effect of thermal and physical properties of the temporal bone on the various quantitative measures that are used clinically for evaluating vestibular function, such as the maximum slow phase speed and its latency. It demonstrates that these measures in post-caloric nystagmus are not only a function of the vestibular system, but to a certain extent, depend on the thermophysical properties of the temporal bone.

References

1   Baertschi, A.; Johnson, R.; Hanna, G.: A theoretical and experimental determination of vestibular dynamics in caloric stimulation. Biol. Cybernetics *20:* 175–186 (1975).
2   Bock, O.; Bromm, B.: A mathematical model of caloric nystagmus. Biol. Cybernetics *27:* 27–32 (1977).
3   Chalabi, Z.: Stochastic signal analysis of thermally induced vestibular electronystagmograms; Thesis London (1981).
4   Demers, R.; Outerbridge, J.S.: Theoretical analysis of vestibular response to caloric stimulation. 5th Can. Med. Biol. Engn Conf. Montreal 1974.
5   Harrington, J.: Caloric stimulation of the labyrinth – experimental observation. Am. Lar. Rhinol. Oto. Soc., Columbus 1969.
6   Steer, R.: The influence of angular and linear acceleration and thermal stimulation on the human semicircular canal; thesis Cambridge, Mass. (1967).
7   Young, J.: Analysis of vestibular system responses to thermal gradients induced in the temporal bone; thesis Ann Arbor (1972).

Z. Chalabi, PhD, Royal National Throat, Nose and Ear Hospital,
London WC1 (England)

Adv. Oto-Rhino-Laryng., vol. 30, pp. 46-49 (Karger, Basel 1983)

Counterdrifting of the Eyes following Unilateral Labyrinthine Disorders

Klaus Hess

Neurological Clinic, University Hospital, Zürich, Switzerland

Introduction

In this paper, a newly observed oculomotor phenomenon accompanying some vestibular disorders is demonstrated. It may help to differentiate between central compensation alone and additional peripheral influences, such as labyrinthine recovery.

Material and Methods

Eye movements and eye positions of 5 patients with acute vestibulopathy and of 5 patients with vestibular neurectomy were monitored during the course of their illness. Vestibular neurectomy including Scarpa's ganglion was performed because of sudden unilateral deafness and persisting tinnitus in 3 patients, Ménière's disease in 1 and acoustic neuroma in another patient. Preoperative calorics were normal in all but the acoustic neuroma patient. – The diagnosis of acute vestibulopathy was based on an episode of acute vertigo lasting for days, in association with the findings of unilateral vestibular failure, as defined by *Drachman and Hart* [1972]. Electrooculograms of horizontal and vertical eye movements were obtained by skin electrodes at the outer canthi of the eyes. Recordings were direct current with a 30 Hz low pass filter. For the examination, patients were seated on a turntable which was encapsulated by a light-proof drum. Each subject's head was positioned in a comfortable head-rest. After calibration, patients were asked to fixate upon a small (diameter 0.8°), dim target light, located straight ahead (0°), which after a few seconds was switched off (fig. 1, 0° ▲). They were instructed to maintain imaginative fixation for some 20 s in complete darkness, and to catch the target point immediately after its reappearance (fig. 1, △) at the same place. Amount of eye deviation could be registered by the amplitude of the necessary visually guided saccade. Thereafter, the same procedure was performed at alternate lateral gaze angles of 20, 30, and 40°, patients being asked not to move their heads.

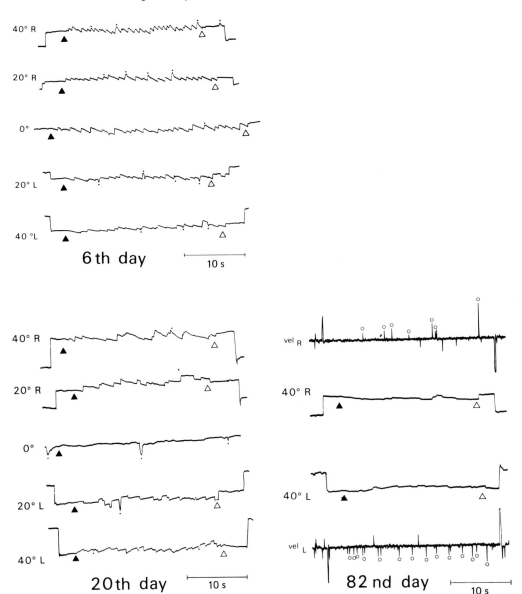

Fig. 1. Acute left-sided vestibulopathy in a 33-year-old patient on the 6th, 20th and 82nd day after onset. Direct current oculography. Calibration: initial and final visually guided horizontal saccades. Initial eye position in each trace is midposition in orbit. R = To the right; L = to the left; ▲ = target light off; △ = target light on; ● = blink; ○ = drift-correcting saccades as indicated by the velocity trace (uncalibrated).

Results

(1) Vestibular neurectomy: Spontaneous nystagmus, which in the more acute stages of the illness was modified according to Alexander's law, slowly declined within a few weeks. The slow phase of nystagmus, i.e. vestibular drifting, was unidirectional (to the same side) in each position of gaze and in each stage (controls up to 20 months p.o.) during the course of the illness.

(2) Acute vestibulopathy: By contrast, in all 5 patients with acute vestibulopathy a clear change of drift direction (fig. 1, 6th and 20th day) was recognizable on lateral gaze to the side opposite of quick phase of spontaneous nystagmus. Direction changing occurred between the 5th and the 20th day after onset of the disease: At the same time, spontaneous nystagmus on direct forward gaze became weak and finally disappeared, yet was still present on lateral gaze to the other side (fig. 1, 20th day). Therefore, the secondary and oppositely directed drift which was called counterdrifting of the eyes ['Gegendrift'; *Hess,* 1981] produced a gaze nystagmus-like pattern of eye movements in the dark. In the subsequent examinations, slow and more or less symmetrical reduction of vestibular drift and counterdrifting was observed (fig. 1, 82nd day). In 1 patient, so-called recovery nystagmus [*Stenger,* 1959] on direct forward gaze developed in the course of intense counterdrifting. Recovery nystagmus persisted for more than 2 months, asymmetric counterdrifting even longer. After recovery, all 5 patients showed bilateral caloric responses. In 3 of them, including the patient with recovery nystagmus, return of function of the initially unresponsive labyrinth was demonstrable.

Discussion

Counterdrifting of the eyes is a poorly understood oculomotor phenomenon which develops by conversion of unidirectional vestibular drift to a bilateral centripetal (gaze nystagmus-like) pattern. It was observed in patients with acute vestibulopathy only, but so far not in patients with vestibular neurectomy. This suggests that some peripheral influence (sensorineural epithelium? Scarpa's ganglion?) is involved in the development of counterdrifting. In contrast to vestibular neurectomy, in acute vestibulopathy both repair (i.e. return to the original function) and compensation (i.e. achievement of function by means of central rearrangement) may interact to various degrees. In 1 patient of this series, counterdrifting was found to be connected with recovery nystagmus [*Stenger,* 1959]. Whether counterdrifting is based on the same mechanism as recovery nystagmus or is different from it, is not yet clear.

References

Drachman, D.A.; Hart, C.W.: An approach to the dizzy patient. Neurology, Minneap. *22:* 323–334 (1972).

Hess, K.: Compensatory eye drifting after 8th nerve lesions. Clin. Neurol. Neurosurg. *83:* 176 (1981).

Stenger, H.H.: 'Erholungsnystagmus' nach einseitigem Vestibularisausfall, ein dem Bechterew-Nystagmus verwandter Vorgang. Archs Oto-Rhino-Lar. *175:* 545–549 (1959).

K. Hess, MD, Neurological Clinic, University Hospital, CH–8091 Zürich (Switzerland)

Adv. Oto-Rhino-Laryng., vol. 30, pp. 50–53 (Karger, Basel 1983)

Labyrinthine Activation during Rotation about Axes Tilted from the Vertical

Theodore Raphan, Walter Waespe, Bernard Cohen

Department of Computer and Information Science, Brooklyn College,
City University of New York, N.Y., USA: Department of Neurology, University
of Zürich, Switzerland, and Department of Neurology, Mount Sinai School of
Medicine, New York, N.Y., USA

Introduction

Changing the orientation of the head with regard to gravity has a dramatic effect on ocular compensatory movements generated by the vestibular system. Rotation about axes tilted from the vertical induces continuous nystagmus for as long as the stimulus persists and cancels the post-rotatory response when rotation stops [1, 4, 6, 7]. Postulates as to the origin of activity responsible for generating the nystagmus during off vertical axis rotation include activation of the otoliths [4] and the semicircular canals [1]. We have recently shown that off vertical axis rotation causes continuous nystagmus primarily by exciting the velocity storage mechanism that generates optokinetic after-nystagmus (OKAN) and is involved in visual-vestibular interactions [2, 5, 6]. However, labyrinthine mechanisms that maintain excitation in the velocity storage integrator are still unclear. In this study we analyzed unit activity recorded in the vestibular nerve of alert monkeys to determine whether semicircular canal and otolith afferents have activity that is related to the slow phase velocity of nystagmus generated during off vertical axis rotation.

Methods

Experiments were performed on Rhesus (*Macaca mulatta*) and Cynomolgous (*Macaca fascicularis*) monkeys. Eye movements were recorded with EOG. The EOG was differentiated and rectified to obtain slow phase velocity. Unit activity was recorded with tungsten microelectrodes from the eighth nerve at its point of entry into the brainstem. During testing monkeys sat in a primate chair that could be rotated about a vertical axis or about axes tilted from the vertical [6]. Average frequency of firing was obtained by a computer program that counted the number of spikes within a 192-ms window und updated the count every 1.6 ms. The algorithm had a resolution of 5 Hz, a frequency response of a zero-order hold system and a linear range up to 625 Hz.

Results

Neurons were tested using velocities from 3.5 to 180°/s and tilt angles of 30, 60, and 90°. For a step in velocity, semicircular canal units had an increase in frequency which decayed to a steady state with a time constant of 3–6 s (fig. 1A) while the nystagmus continued. These units did not show a direction specific, steady state response to off vertical axis rotation [3]. That is, the steady state average frequency of firing was not affected by the velocity or direction of rotation. Firing rates were either the same as the spontaneous levels or increased slightly. In addition there was little or no modulation in relation to head position. This shows that the continuous velocity signal responsible for the nystagmus is not emanating directly from the semicircular canals.

Otolith units on the other hand modulated their activity in relation to head position. The firing rates of 'regular' otolith neurons were approximately sinusoidal and the peak firing rates occurred at head positions that were approximately independent of the velocity of rotation, direction of rotation or angle of tilt (fig. 1B). The depth of modulation was related to tilt angle and presumably was dependent on the neuron's polarization angle. Firing rates of 'irregular' neurons, including those that had no steady state response to static head tilts, were also modulated sinusoidally. The peak frequency for a given neuron occurred at a specific head position while rotating. There were phase differences for oppositely directed rotations. However, the phase of the peak frequency was not significantly altered with speed of rotation or angle of tilt as long as the rotation was unidirectional. Both 'regular' as well as 'irregular' neurons had phase distributions over a 360° range. This is consistent with the polarization angles associated with otolith afferents. The average frequency of the sinusoidal modulation in unit activity was approximately independent of tilt angle or direction of rotation. The average frequency of the sinusoidal modulaton increased somewhat with stimulus velocity. However, there was no difference for

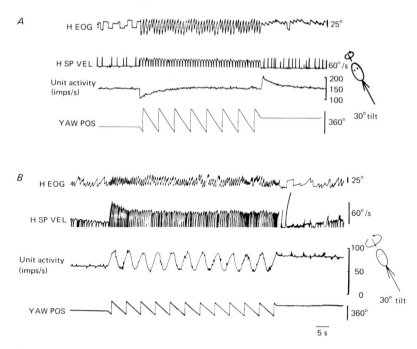

Fig. 1. Response of a 'regular' horizontal semicircular canal afferent (*A*) and a 're-gular' otolith afferent (*B*) to a step in head velocity during off vertical axis rotation. The top traces in *A* and *B* are the horizontal EOG, the second traces show the velocity of the nystagmus, the third traces show the frequency of firing of the units and the bottom traces show the angular position of the head. Note the decay of the frequency of firing of the canal afferent to its spontaneous level while the nystagmus continues (*A*). The otolith afferent is continuously modulated with peaks occurring at a specific head position (*B*).

rotation to the right or left. Therefore, signals coming from the otolith organs are not simply related to the velocity of the continuous nystagmus during off vertical axis rotation.

Conclusion

The data indicate that neither the semicircular canals nor the otolith organs provide a signal that is directly responsible for the generation of continuous nystagmus during off vertical axis rotation. The semicircular canals play a subordinate role. The unchanging phase relation of peak frequency relative to a given head position for 'regular' as well as 'irregular' otolith units during unidirectional rotation is consistent with the theory that sequential activation of otolith units induces a traveling wave pattern in the cellular structure of the maculae. It is postulated that the velocity of this wave is detected centrally and converted to a signal that activates the velocity storage integrator which in turn generates the continuous nystagmus [6].

References

1   Benson, A.J.; Bodin, M.A.: Aerospace Med. *37:* 144–154 (1966).
2   Cohen, B.; Matsuo, V.; Raphan, T.: J. Physiol., Lond. *270:* 321–344 (1977).
3   Goldberg, J.M.; Fernandez, C.: Ann. N.Y. Acad. Sci. *374:* 40–43 (1981).
4   Guedry, F.E., Jr.: Acta oto-lar. *60:* 30–48 (1965).
5   Raphan, T.; Matsuo, V.; Cohen, B.: Exp. Brain Res. *35:* 229–248 (1979).
6   Raphan, T.; Cohen, B.; Henn, V.: Ann. N.Y. Acad. Sci. *374:* 44–55 (1981).
7   Young, L.R.; Henn, V.: Fortschr., Zool. *23:* 235–246 (1975).

T. Raphan, PhD, Department of Computer and Information Science,
Brooklyn College, City University of New York, Bedford Avenue and Avenue H,
Brooklyn, NY 11210 (USA)

Adv. Oto-Rhino-Laryng., vol. 30, pp. 54–57 (Karger, Basel 1983)

Vestibular Evoked Potentials in the Awake Rhesus Monkey

Andreas Böhmer, Volker Henn, Dietrich Lehmann

Neurological Department, University Hospital, Zürich, Switzerland

Averaging scalp-recorded cortical and brainstem potentials, evoked by repetitive application of adequate stimuli, is a well-established tool in clinical diagnosis of neurological and otological diseases. The method requires stimuli with a fast rising time (step functions of light, touch, or sound intensity) which for the vestibular system are difficult to produce. In spite of a few attempts [*Spiegel* et al., 1968; *Schmidt*, 1979; *Elidan* et al., 1982] up to now, no clinically applicable method has been established, and clinical vestibular examination still is restricted to measurements of compensatory eye movements as one main output of the system.

We present here preliminary data of an attempt to record short latency vestibular potentials in awake Rhesus monkeys, using a primate rotating chair providing short angular accelerations of 2,000 deg/s^2. To avoid muscle artefact problems EEG electrodes were implanted deeply into the animal's bony skull.

The data were obtained from 1 normal Rhesus monkey, 1 labyrinthectomized monkey, and 1 animal in which both horizontal semicircular canal nerves were cut. The monkeys were chronically prepared with head bolts to fixate the head during experiments. Silver-silver chloride electrodes were implanted around the bony orbit to measure eye position and at different sites in the bony skull for EEG recordings. In this report, we will describe the evoked potential recordings between either mastoid and a midline site over the occiput. During the recording the monkey sat upright on the torque motor-driven rotating chair which rotated either the whole animal, or the head only with the trunk

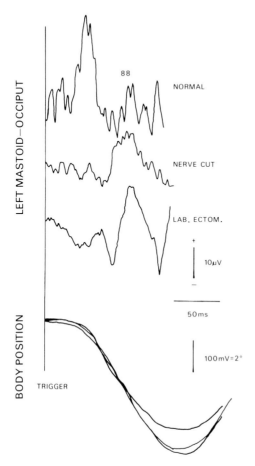

LEFT MASTOID–OCCIPUT

88

NORMAL

NERVE CUT

LAB. ECTOM.

10μV

50ms

100mV = 2°

BODY POSITION

TRIGGER

Fig. 1. Averaged (n = 60) EEG potentials recorded from left mastoid versus occiput in 3 different monkeys, evoked by a counter-clockwise sinusoid acceleration of the whole animal. At bottom the position curves of the three corresponding trials are superimposed.

immobilized, or the trunk only with the head immobilized. A fast, sinusoidal movement of 8 deg amplitude and about 250 ms duration was used as a stimulus (fig. 1, bottom). For each trial about 120 such stimuli were applied in complete darkness. EOG, EEG, and body position signals were amplified, filtered, and stored on analog tape. These signals were averaged (n = 30, 60, or 90) off-line, using a CAT 1000 hardwired averaging device.

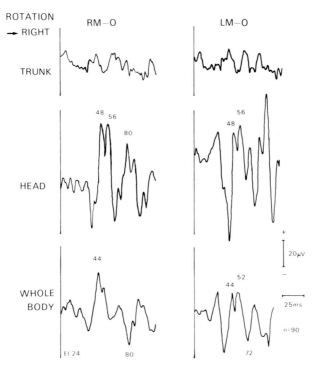

Fig. 2. Averaged (n = 90) evoked EEG potentials in the normal monkey recorded from right (RM-O) and left (LM-O) mastoid versus occiput in response to clockwise trunk, head, and whole body acceleration.

Figure 1 shows such averaged evoked potentials – left mastoid versus occiput recordings – from our 3 monkeys. In the normal monkey a large, mastoid positive potential appears 44 ms after the trigger. This peak is absent in the 2 monkeys with nonfunctioning horizontal vestibular systems. A second mastoid positive peak appears 40 ms later, but is present also in the control animals and thus is not a vestibular response. In our paradigm it is difficult to determine real latencies because there is no sharp onset of accleration, but body movement starts gradually some 30 ms after the trigger which initiates the turntable motion. Therefore all latencies are related to the trigger. The first peak is well reproducable in the normal monkey with different stimulation conditions and in recordings from both sides (fig. 2). Rotation of only the head results in latencies 4 ms longer due to the inertia of the stimulating device; the same latency differences were also found in the chair position signals.

To prove that these potentials are caused by vestibular activation, the following other possible sources were excluded: eye movement signals, neck muscle potentials, auditory evoked potentials induced by the noise of the rotating chair, and electromechanical artefacts, e.g. electrical field changes of the chair motor. The latter as well as neck muscle potentials may be excluded by the flat traces obtained by trunk rotation only, using the same motor (fig. 2, top row); this excludes also to some extent auditory evoked potentials. At least in the control animal with the nerves cut, hearing was intact, and this monkey showed no peak at 44 ms (fig. 1). In the labyrinthectomized animal cervico-ocular reflexes were increased. Thus, eye movements in response to head rotation were quite similar to those of the normal monkey. In both cases the horizontal EOG represented a mirror image of the head position signal, without the early peaks of the evoked potential.

Summing up, we have demonstrated a short latency evoked potential to vestibular stimuli in monkeys. As we have succeeded also in preliminary tests in these animals to obtain such potentials, using a conventional human rotating chair with acceleration limited to $750°/s^2$, we hope it will be possible to adapt this method for vestibular diagnosis in human patients.

References

Elidan, J.; Sohmer, H.; Nizan, M.: Recording of short latency vestibular evoked potentials to acceleration in rats by means of skin electrodes. EEG clin. Neurophysiol. *53:* 501–505 (1982).

Schmidt, C.L.: Elektrisch evozierte, computer-gemittelte vestibuläre Hirnstammpotentiale bei der Katze. Archs Otolar. *222:* 199–204 (1979).

Spiegel, E.A.; Szekely, E.G.; Moffet, R.: Cortical responses to rotation. Acta oto-lar. *66:* 81–88 (1968).

A. Böhmer, MD, Neurological Department, University Hospital,
CH-8091 Zürich (Switzerland)

Adv. Oto-Rhino-Laryng., vol. 30, pp. 58–63 (Karger, Basel 1983)

Possible Neurotransmitters Involved in Excitatory and Inhibitory Effects from Inferior Olive to Contralateral Lateral Vestibular Nucleus

I. Matsuoka[a], *J. Ito*[a], *M. Sasa*[b], *S. Takaori*[b]

[a] Department of Otorhinolaryngology and [b] Department of Pharmacology, Faculty of Medicine, Kyoto University, Kyoto, Japan

The inferior olive (IO) sends climbing fibers to the cerebellar Purkinje cells, which in turn produce an inhibition of the lateral vestibular nucleus (LVN) neurons [5, 7, 12]. *Brodal* [4] first found a fiber connection from the IO to the contralateral LVN, using a degeneration method. Recently, we have confirmed the existence of direct projection to the LVN from the dorsal cap and β-nucleus of the contralateral IO, using a retrograde horseradish peroxidase (HRP) tracing technique [6]. Thus, an electrophysiological study was performed to elucidate the role of the IO on the LVN neurons and the neurotransmitters involved in this pathway, using cats in which the cerebellum had been removed to avoid any effects from the cerebellar Purkinje cells on the LVN neurons.

Methods

20 adult cats weighing 2.5–3.5 kg were anesthetized initially with diethyl ether and then α-chloralose (30 mg/kg, i.v.). A bipolar stainless steel electrode for stimulation of the vestibular nerve was inserted through the round window of the middle ear cavity. After fixation of the animal's head in a stereotaxic instrument, the cerebellum was totally removed by aspiration. The animal was then immobilized with gallamine triethiodide (5 mg/kg/h, i.v.) under artificial respiration, and body temperature was maintained at 36.5–37.5 °C with a heating pad. Extracellular neuron activity in the left LVN (P: 8.0, L: 4.0, H: –3.5 to –4.5) [14] was recorded using a glass-insulated silver wire microelectrode (electrical resistance: approximately 1 MΩ) attached along a seven-barreled micropipette, which had a tip diameter of 7–10 μm, and was filled with 0.2 M atropine sulfate, 5 mM bicuculline hydrochloride, 1.0 M GABA, 10 mM strychnine sulfate, 0.5 M glycine,

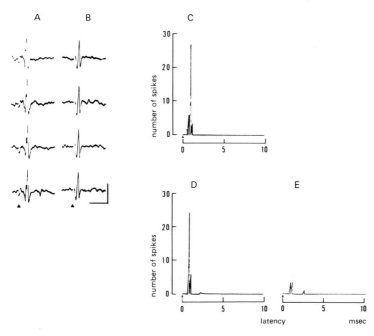

Fig. 1. Spike generation in lateral vestibular nucleus neuron upon vestibular nerve (*A*) and contralateral inferior olive (IO) stimulation (*B*), and effects of microiontophoretically applied atropine on the spikes produced by IO stimulation (*C–E*). *C–E* Poststimulus latency histograms obtained from 40 successive responses of controls (*C*) and during application of atropine of 100 (*D*) and 200 nA (*E*). ▲ = Stimulus artifacts applied to the vestibular nerve (*A*) and IO (*B*). Calibration: 5 ms and 0.5 mV.

0.2 *M* diphenhydramine hydrochloride and 1.0 *M* monosodium glutamate. These chemicals were microiontophoretically applied to the immediate vicinity of the neuron using a Microiontophoresis Programmer (WP-I, Model 160); a retaining current of 10–25 nA was used during the recording. A bipolar stainless steel stimulating electrode with a tip diameter of 0.2 mm and tip separation of 0.2 mm was inserted into the contralateral dorsal cap of the IO (P: 10.0, L: 1.5, H: –10.0). Test stimuli, composed of a square pulse (0.05 ms and 1–10 V), were applied to the vestibular nerve and the IO, every 1.6 s. A conditioning stimulus (square pulse, 0.05 ms and 1–10 V) was given to the IO at various intervals preceding the test stimulus to the vestibular nerve (C-T interval). At least 10 successive responses were amplified and displayed on an oscilloscope (Nihon Kohden, VC-9), and photographed. The statistical significance of the data was determined by Student's t test. After the termination of each experiment, the stimulated sites of the IO were marked by passing a cathodal current of 20 μA for 10–20 s, and then checked by staining with cresyl violet. Other details of the experiment are outlined in previous papers [6, 9, 13].

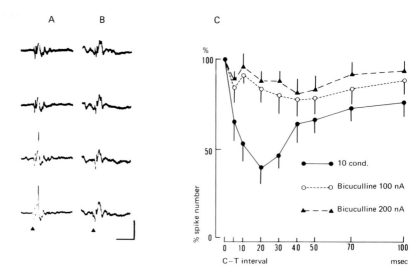

Fig. 2. Inhibitory effects of the conditioning stimulus applied to the contralateral inferior olive (IO) on spike generation with vestibular nerve stimulation, and antagonizing effect of bicuculline on the IO-induced inhibition. A Spikes produced by vestibular nerve stimulation (▲). B Inhibition by IO conditioning stimulation of the spike generation upon vestibular nerve stimulation. C Effects of the time intervals between conditioning and testing stimuli (C-T interval) on the vestibular nerve-induced spike, and effects of microiontophoretic application of bicuculline. Vertical bar of each point indicates standard error (n = 7). Calibration: 5 ms and 0.5 mV.

Results

The effects of IO stimulation were examined on the LVN neuron which was monosynaptically activated by vestibular nerve stimulation. Out of 168 monosynaptic LVN neurons, 43 neurons were also activated by the IO stimulation; a typical example is shown in figure 1A and B. The mean spike latencies of these neurons with vestibular nerve and IO stimulation were 1.42 ± 0.07 and 1.40 ± 0.08 ms, respectively. When atropine was microiontophoretically applied, the spike generation of the LVN neuron upon IO stimulation was dose-dependently inhibited (fig. 1C–E). The mean spike numbers of 16 neurons upon the IO stimulation was significantly (p <0.01) reduced from 0.94 ± 0.03 to 0.63 ± 0.08 and 0.28 ± 0.09 with atropine of 100 and 200 nA, respectively. In 34 monosynaptic LVN neurons, the conditioning stimulus applied to the IO preceding the test stimulus to the vestibular nerve produced an inhibition of spike generation upon stimulation (fig. 2A, B). When the conditioning stimulus preceded the test stimulus by 30 ms, the mean spike number of 22 neurons was significantly reduced from 1.43 ± 0.11 to 0.46 ± 0.09 (p <0.01). The inhibition took place within 5 ms of the C-T interval and persisted for 50 ms (fig. 2C). When bicuculline, which is a GABA antagonist, was microiontophoretically applied, the inhibition was blocked (fig. 2C). In the presence of an IO conditioning stimulation 20 ms preceding the test stimulus, the mean spike number of 7 neurons was significantly (p <0.01) increased to 1.45 ± 0.22 from 0.68 ± 0.17 by 100 nA of bicuculline. However, microiontophoretically applied strychnine, which is a glycine antagonist, and diphenhydramine up to 100 nA did not block the inhibition by IO conditioning stimulation. Furthermore, microiontophoretic application of GABA (50–100 nA) blocked the spike generation upon vestibular nerve stimulation, as expected, however, glycine up to 200 nA did not significantly affect the spike firing.

Either an excitatory or inhibitory effect upon IO stimulation on the LVN monosynaptic neurons was obtained when the stimulating electrode was positioned in the dorsal cap and β-nucleus of the IO. No response to IO stimulation was observed when the electrode tip was located in other regions in the IO or outside.

Discussion

These histological results coincide with those of our previous HRP studies [6], which have shown direct projection from the dorsal cap and β-nucleus of the IO to the contralateral LVN. Since stimulation of the dorsal cap or β-nucleus of the IO produced spike generation in the LVN neurons with a fairly consistent and relatively short latency of 1.4 ms, the spike was considered to be monosynaptically elicited by IO stimulation. *Allen* et al. [1] have reported that IO stimulation produced an EPSP-IPSP sequence in the contralateral LVN neurons, although they have not specified the stimulated sites in the IO. They have concluded that the EPSP is monosynaptically produced through collaterals of the climbing fibers originating in the IO and the IPSP via the cerebellar Purkinje cells. Our findings that the LVN monosynaptic neurons were also monosynaptically activated by stimulation of the dorsal cap and β-nucleus of the IO, are in accordance with those of *Allen* et al. [1], although it cannot be concluded whether or not the spikes in our study, obtained upon IO stimulation, were induced through the activation of the collaterals of climbing fibers. Since the monosynaptic spike upon IO stimulation was dose-dependently inhibited by microiontophoretic application of atropine, the neurotransmitter involved in this pathway appears to be acetylcholine. This possibility is further supported by the immunohistochemical findings of *Kimura* et al. [8] that there are cells and terminals containing acetylcholinetransferase in the IO and LVN, respectively. The inhibition by IO conditioning stimulation of the LVN monosynaptic neurons is not due to the activation of the cerebellar Purkinje cells, because the effects were obtained in the decerebellated animals. Although the inhibition took place within 5 ms after IO stimulation, it cannot be determined at present whether monosynaptically or polysynaptically the inhibition was produced by IO stimulation. However, the findings that bicuculline antagonized the inhibition support the possibility that the inhibition is mediated by GABA at the terminal synapse on the LVN.

The IO, in particular the caudal part of the dorsal cap, receives input from the retina via the lateral tegmental area [2, 11] and from the spinal cord [3, 10]. In addition to the control from the cerebellum of the LVN neurons, our results suggest that the IO may play an important role in integrating the information from the visual and proprioceptive systems and relaying it to the LVN.

References

1   Allen, G.I.; Sabah, N.H.; Toyama, K.: Synaptic actions of peripheral nerve impulses upon Deiters neurones via the climbing fibre afferents. J. Physiol., Lond. *226:* 311–333 (1972).
2   Alley, K.; Baker, R.; Simpson, J.I.: Afferents to the vestibulo-cerebellum and the origin of the visual climbing fibers in the rabbit. Brain Res. *98:* 582–589 (1975).
3   Boesten, A.J.P.; Voogd, J.: Projections of the dorsal column nuclei and the spinal cord on the inferior olive in the cat. J. comp. Neurol. *161:* 215–238 (1977).
4   Brodal, A.: Experimentelle Untersuchungen über die olivocerebellare Lokalisation. Z. ges. Neurol. Psychiat. *169:* 1–153 (1940).
5   Desclin, J.C.: Histological evidence supporting the inferior olive as the major source of cerebellar climbing fibers in the rat. Brain Res. *77:* 365–384 (1974).
6   Ito, J.; Sasa, M.; Matsuoka, I.; Takaori, S.: Afferent projection from reticular nuclei, inferior olive and cerebellum to lateral vestibular nucleus of the cat as demonstrated by horseradish peroxidase. Brain Res. *231:* 427–432 (1982).

7   Ito, M.; Yoshida, M.: The origin of cerebellar-induced inhibition of Deiters neurones. I. Monosynaptic inhibition of the inhibitory postsynaptic potentials. Exp. Brain Res. *2:* 330–349 (1966).
8   Kimura, H.; McGeer, P.L.; Peng, J.H.; McGeer, E.G.: The central cholinergic system studies by choline acetyltransferase immunohistochemistry in the cat. J. comp. Neurol. *200:* 151–201 (1981).
9   Matsuoka, I.; Ito, J.; Sasa, M.; Takaori, S.; Morimoto, M.: Neuronal interaction between ipsilateral medial and lateral vestibular nuclei. Ann. N.Y. Acad. Sci. *374:* 93–101 (1981).
10 Mizuno, N.: An experimental study of the spino-olivary fibers in the rabbit and the cat. J. comp. Neurol. *127:* 267–292 (1966).
11 Mizuno, N.; Mochizuki, K.; Akimoto, C.; Matsushima, R.: Pretectal projections to the inferior olive in the rabbit. Expl. Neurol. *39:* 498–506 (1973).
12 Obata, K.; Ito, M.; Ochi, R.; Sato, N.: Pharmacological properties of the postsynaptic inhibition by Purkinje cell axons and the action of γ-aminobutyric acid on Deiters neurones. Exp. Brain Res. *4:* 43–57 (1967).
13 Sasa, M.; Fijimoto, S.; Igarashi, S.; Munekiyo, K.; Takaori, S.: Microintophoretic studies on noradrenergic inhibition from locus coeruleus of spinal trigeminal nucleus neurons. J. Pharmac. exp. Ther. *210:* 311–315 (1979).
14 Snider, R.S.; Niemer, W.T.: A stereotaxic atlas of the cat brain (University of Chicago Press, Chicago 1961).

I. Matsuoka, MD, Department of Otorhinolaryngology, Faculty of Medicine, Kyoto University, Kyoto 606 (Japan)

Adv. Oto-Rhino-Laryng., vol. 30, pp. 64–70 (Karger, Basel 1983)

Input to Lateral Vestibular Nucleus as Revealed by Retrograde Horseradish Peroxidase Technique

J. Ito[a], *I. Matsuoka*[a], *M. Sasa*[b], *S. Takaori*[b], *M. Morimoto*[c]

Departments of [a] Otorhinolaryngology and [b] Pharmacology, Faculty of Medicine, Kyoto University, Kyoto, Japan; [c] Faculty of Medicine, Kochi Medical School, Kochi, Japan

It has been well documented that the lateral vestibular nucleus (LVN) receives primary afferent fibers from the peripheral vestibular organs, cerebellar nuclei and Purkinje cells of the cerebellum. In addition, our previous study using a retrograde transport technique with horseradish peroxidase (HRP) has demonstrated that there are afferent projections to the LVN from the lateral reticular and gigantocellular nuclei [5]. The existence of innervation from the medial vestibular nucleus (MVN) to the ipsilateral LVN has been previously demonstrated using the HRP technique and electrophysiological method [7]. However, there is still little information concerning connection between the LVN and other vestibular nuclei, and the commissural connection between the LVN and contralateral vestibular nuclei, although main commissural connections exist between the bilateral superior as well as bilateral inferior vestibular nuclei (IVN), and minor commissural connections to the LVN from the contralateral LVN and IVN [2, 6]. Thus, this study was performed to investigate the input system to the LVN from the bilateral vestibular nuclei, reticular nuclei and several parts of cerebellum using a retrograde transport technique with HRP microiontophoretically applied to a highly restricted area in the LVN.

Materials and Methods

10 adult cats weighing 2.5–4.0 kg under anesthesia with α-chloralose (30 mg/kg, i.v.) were used. Schematic representation of the experimental procedure is illustrated in figure 1. A bipolar stainless steel stimulating electrode was inserted into the vestibular

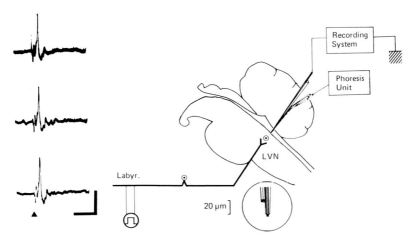

Fig. 1. Schematic representation of the experimental procedure. A spike was monosynaptically elicited by stimulating the vestibular nerve (▲). Calibration: 4 ms, 0.5 mV.

nerve through the round window of the middle ear cavity. Single neuron activity was recorded in the LVN (P: 8.0, L: 4.0, H: −3.5 to −4.5) [10] using a glass-insulated silver wire microelectrode attached along a seven-barrelled micropipette, which had a tip diameter of approximately 1.5 μm and was filled with 1.0 M monosodium glutamate, 10% HRP in Tris-HCl buffer (pH 8.6) and 3.0 M NaCl for balancing. The distance between the tips of the recording microelectrode and micropipette was within 20 μm. After identifying the LVN neuron by stimulating the vestibular nerve and confirming that microiontophoretically applied glutamate increased the firing of the neuron, HRP was ejected to the vicinity of the target neuron by passing an anodal current of 300–500 nA for 5–10 min using a Microiontophoresis Programmer (WP-I, Model 160). After a survival time of 24–36 h, the animal was perfused through the heart with 7% formaldehyde in 0.1 M phosphate buffer. The removed brain was stored in 0.1 M phosphate buffer containing 30% sucrose for 48–72 h, and cut into serial frontal section (50 μm). Each section was treated with benzidine or tetramethylbenzidine in the presence of hydrogen peroxide and counterstained with neutral red. The other details of experimental procedures are similar to those described previously [5, 7, 8].

Results

When the monosynaptic spike of the LVN (fig. 1) was obtained upon vestibular nerve stimulation, HRP was microiontophoretically applied there. Figure 2 shows the HRP injection site, which was highly localized in the LVN to within an area 0.5 mm in diameter.

When the HRP was applied to the ventral or middle part of the

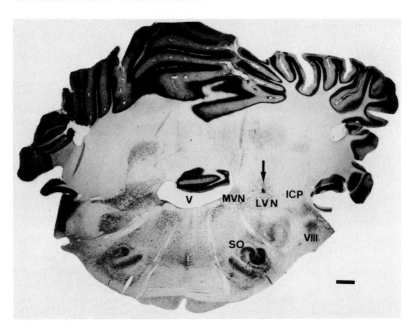

Fig. 2. HRP injection site in the lateral vestibular nucleus (LVN = arrow). ICP = Inferior cerebellar peduncle; MVN = medial vestibular nucleus; SO = superior olive nucleus; V = fourth ventricle; VIII = VIIIth cranial nerve. Bar = 1.0 mm.

LVN, HRP-reactive cells were found in the bilateral MVN, bilateral IVN and contralateral LVN (fig. 3, 4). The HRP-reactive cells in the ipsilateral MVN and IVN were observed in the caudoventral and rostroventral parts of the nuclei, respectively. These cells were multipolar and medium-sized. The HRP-positive cells found in the contralateral LVN were mainly located in the ventral part, although the number of the cells labelled with HRP was smaller than those in the MVN and IVN. The HRP-reactive cells in the LVN included small to medium-sized cells, but none of the giant cells contained HRP granule. In the reticular formation, there was a considerable number of HRP-reactive cells in the ipsilateral lateral reticular nucleus at the caudal level of the ventral cochlear nucleus (fig. 3, 4) and a few cells in the bilateral gigantocellular nucleus at the level of facial nucleus. The HRP-reactive cells in the lateral reticular nucleus were medium-sized with a multipolar shape; in the gigantocellular nucleus they were small to medium-sized and none of the giant cells reacted with HRP. The HRP-positive cells

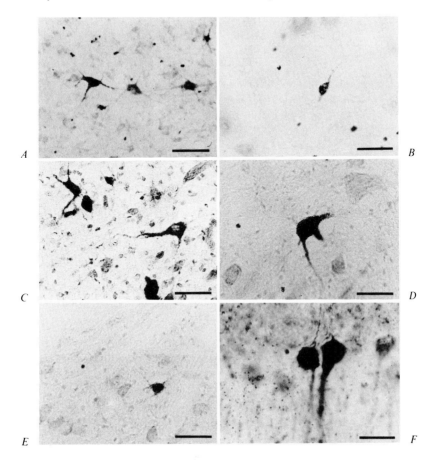

Fig. 3. HRP-labelled cells. *A* Ipsilateral medial vestibular nucleus. *B* Ipsilateral inferior vestibular nucleus. *C* Contralateral medial vestibular nucleus. *D* Ipsilateral lateral reticular nucleus. *E* Contralateral inferior olive nucleus. *F* Ipsilateral nodulus. Bars = 50 μm.

were also found in the contralateral inferior olive, mainly in the dorsal cap and β-nucleus (fig. 3). In the cerebellum, many HRP-positive cells were observed in the ipsilateral rostral part and contralateral caudal part of the fastigial nucleus; a few cells were found in the ipsilateral rostral part of interpositus nucleus as well as in the bilateral rostroventral area of the dentate nuclei (fig. 3, 4). HRP-labelled Purkinje cells were located in the ipsilateral cerebellar nodulus, lingula and flocculus (fig. 3).

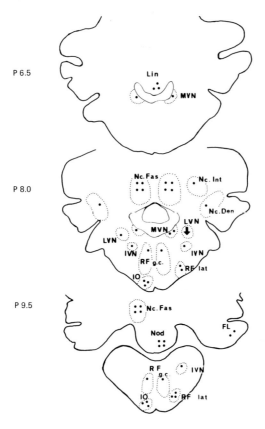

Fig. 4. Schematic representation of the location of HRP-positive cells. Each dot represents the relative number of cells containing HRP granule. FL = Flocculus; IO = inferior olive; IVN = inferior vestibular nucleus; Lin = lingula; LVN = lateral vestibular
nucleus; MVN = medial vestibular nucleus; Nc. den = dentate nucleus; Nc. fas = fastigial nucleus; Nc. int = interpositus nucleus; Nod = nodulus; RF g.c. = gigantocellular
nucleus; RF lat = lateral reticular nucleus.

Discussion

A highly restricted HRP deposit has been reportedly obtained with
only minimal damage to the injection site using a microiontophoretic
technique [3, 4]. Our HRP deposit was limited to an area within 0.5 mm
in diameter and, therefore, could be localized in either the dorsal or
ventral part of the LVN. Under these conditions, the present study confirmed our earlier results that the LVN receives innervation from the

bilateral fastigial nuclei, bilateral dentate nuclei and ipsilateral interpositus nucleus of the cerebellum and from Purkinje cells in the nodulus, lingula and flocculus [5]. These results are in good agreement with those obtained in degeneration studies [1, 2, 11] except for the dentate and interpositus nuclei. In addition to input to the LVN from the inferior olivary nucleus, reticular input to the LVN from the ipsilateral lateral reticular and bilateral gigantocellular nuclei was also confirmed in the present HRP study. Furthermore, when the HRP deposit was localized in the middle of the LVN, HRP-reactive cells were also found in the ipsilateral MVN, reconfirming the previous findings that there are direct connections from the MVN to the LVN [7]. At the same time, it was found that the LVN received direct input from the contralateral MVN, bilateral IVN and contralateral LVN.

Commissural connections of vestibular nuclei have been extensively studied by *Ladpli and Brodal* [6] using the Nauta method. They have reported that the superior vestibular nucleus and IVN send fibers to the contralateral superior vestibular nucleus and IVN, respectively. Degeneration study has also demonstrated that the LVN receives minor commissural fibers from the LVN and IVN [2, 6]. In our HRP study, the cells in the LVN projecting to the contralateral LVN were few in number, but those in the MVN and IVN were considerable. It is unlikely that the HRP-reactive cells found in the contralateral MVN resulted from the uptake of HRP from the fiber passing through the immediate ventral area of the LVN, because the HRP-positive cells were obtained even when the HRP was applied to the middle of the LVN far from the passing fibers. It is of interest that the giant cells in the LVN and IVN did not project to the contralateral LVN. Since the electrophysiological studies have demonstrated that the commissural fibers produce an inhibition of ventral neurons in the vestibular nuclei [9], it is likely that small to medium-sized cells in these nuclei are involved in the inhibitory mechanism.

References

1   Angaut, P.; Brodal, A.: The projection of the 'vestibulo-cerebellum' onto the vestibular nuclei in the cat. Archs ital. Biol. *105:* 441–479 (1967).
2   Brodal, A.: Anatomy of the vestibular nuclei and their connections; in Handbook of sensory physiology, vol. VI, pp. 240–352 (Springer, Berlin 1974).
3   Gallager, D.W.; Pert, A.: Afferents to brain stem nuclei (brain stem raphe, nucleus

reticularis pontis caudalis and nucleus gigantocellularis) in the rat as demonstrated by microiontophoretically applied horseradish peroxidase. Brain Res. *144:*257-275 (1978).

4   Graybiel, A.M.; Devor, M.: A microelectrophoretic delivery technique for use with horseradish peroxidase. Brain Res. *68:* 167-173 (1974).

5   Ito, J.; Sasa, M.; Matsuoka, I.; Takaori, S.: Afferent projection from reticular nuclei, inferior olive and cerebellum to lateral vestibular nucleus of the cat as demonstrated by horseradish peroxidase. Brain Res. *231:* 427-432 (1982).

6   Ladpli, R.; Brodal, A.: Experimental studies of commissural and reticular formation projections from the vestibular nuclei in the cat. Brain Res. *8:* 65-96 (1968).

7   Matsuoka, I.; Ito, J.; Sasa, M.; Takaori, S.; Morimoto, M.: Neuronal interaction between ipsilateral medial and lateral vestibular nuclei. Ann. N. Y. Acad. Sci. *374:* 93-101 (1981).

8   Sasa, M.; Fujimoto, S.; Igarashi, S.; Munekiyo, K.; Takaori, S.: Microiontophoretic studies on noradrenergic inhibition from locus coeruleus of spinal trigeminal nucleus neurons. J. Pharmac. exp. Ther. *210:* 311-315 (1979).

9   Shimazu, H.; Precht, W.: Inhibition of central vestibular neurons from the contralateral labyrinth and its mediating pathway. J. Neurophysiol. *29:* 467-492 (1966).

10 Snider, R.S.; Niemer, W.T.: A stereotaxic atlas of the cat brain (University of Chicago Press, Chicago 1961).

11 Walberg, F.; Jansen, J.: Cerebellar corticovestibular fibers in the cat. Expl Neurol. *3:* 32-52 (1961).

J. Ito, MD, Department of Otorhinolaryngology, Faculty of Medicine,
Kyoto University, Kyoto 606 (Japan)

Adv. Oto-Rhino-Laryng., vol. 30, pp. 71–75 (Karger, Basel 1983)

The Saccade Spike

Ville Jäntti[a], *Eero Aantaa*[b] *Heikki Lang*[c], *Lucyna Schalén*[d], *Ilmari Pyykkö*[e, 1]

Departments of [a] Pharmacology, [b] Otorhinolaryngology, and [c] Clinical Neurophysiology, University Hospital of Turku, Finland; [d] Balanslaboratoriet, Lund; [e] Department of Otorhinolaryngology, University Hospital of Lund, Sweden

Introduction

The saccade spike is a spike-like artefact that can be seen in electrooculographic (EOG) recordings in the beginning of saccades. The saccade spike is often polyphasic with a negative main component [*Jäntti*, 1982]. The duration of the spike is 10–30 ms and its amplitude can exceed 100 μV. The spike that is seen in EOG is composed of two individual spikes, one in each recording electrode. In binocular recording the difference of the spikes is recorded by the two electrodes [*Jäntti*, 1982].

The significance of the saccade spike as a possible source of error in the interpretation of EOG is not commonly noticed. In this review we demonstrate the occurrence of the spike and the variation in the spike depending on the electrode position by recording the saccadic eye movements.

Methods

The EOG tracings were recorded binocularly with the Siemens-Elema ink-jet recorder. Various low-pass filters from 15 to 500 Hz were used and either a DC recording or a

[1] We thank *Pekka Airikkala*, MD, for assistance in preparing figure 3 and Asst. Prof. *N.G. Henriksson*, MD, for valuable discussions. This study was supported financially by the Paulo Foundation, Finland.

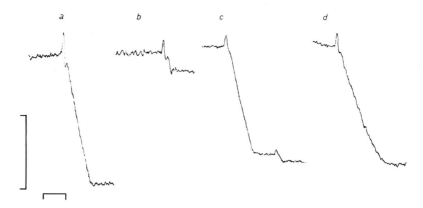

Fig. 1. a 40° saccade. *b* 5° *saccade. c* Saccade with a corrective saccade. *d* Saccade made with eyes closed. All these saccades are from the same test person. DC recording, low-pass filter at 70 Hz in *a, b* and *d* and 15 Hz in *c.* Calibrations: vertical bar 200 μV, horizontal bar 0.1 s.

0.5-Hz high-pass filter were employed. The paper speed was 100 mm/s. the saccade spikes occurring in connection with the rapid eye movements of REM stage of sleep were recorded with the same type of ink-jet recorder with paper speed of 1 cm/s.

Disposable Ag-AgCl electrodes (Medicotest A-15-VS) with an electrode paste cup of 7 mm in diameter were used in the recording of saccades. The distance from the electrode to the outer canthi was 15 mm. To get closer to the canthi other electrodes (tin cups of 7 mm in diameter) were used, which were bonded beside the outer canthi.

Results

Saccadic spikes at different amplitudes of saccades are exhibited in figures 1 and 2. In general, larger saccades were connected with larger spikes, but the smaller saccades were relatively more disturbed by the spikes. The spikes occurred in various degrees in different subjects and also within one subject between different saccades. Noteworthy is that saccade spikes could be observed in all kinds of saccades including corrective saccades (fig. 1c), saccades made behind closed eyelids (fig. 1d) as well as in the rapid eye movements during the REM stage of sleep (fig. 3).

Filtering of the EOG signal does not totally eliminate the spikes from the saccades. If the spike is prominent in the EOG signal, it is still

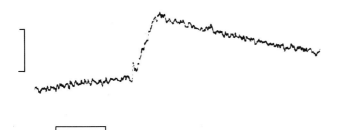

Fig. 2. A 10° saccade of the same test subject as in figure 1. AC recording, frequency limits 0.5–500 Hz. Electrodes are placed just at the outer canthi.

very distinct after filtering with 15 Hz low-pass filter (fig. 1c). The interference of the spike with saccades increases by using higher frequency limits (70 Hz, fig. 1a, b).

The position of the electrodes is critical in recording of the saccades as well as the spikes. In the close vicinity of the canthi of the eyes the amplitude of the saccade is higher, absolute and in relation to the spike (fig. 3) than when the electrodes are placed further away from the canthi (fig. 1).

Discussion

The saccade spike was first described by *Blinn* [1955] who thought that the spike originated from the lateral rectus muscle of the eye; thus he named it 'the external rectus spike'. Later, however, it has become evident that it originates from the muscles of the face [*Beaussart and Guieu*, 1977].

Patients or normal test subjects often slightly blink during saccades and fast phases of nystagmus; sometimes only a slight movement of the lower eyelid is observed. On the other hand, especially in EEG recordings a small spike is often seen in the beginning of the blink artefact. It seems likely, therefore, that the saccade spike is connected with the co-activation of the orbicularis oculi muscle. It might, however, also be partly generated by other muscles of the face. The retroauricular muscle, for instance, is activated by lateral gaze and in some animals the ears move synchronously with the eyes during nystagmus [*Schmidt and Thoden*, 1978].

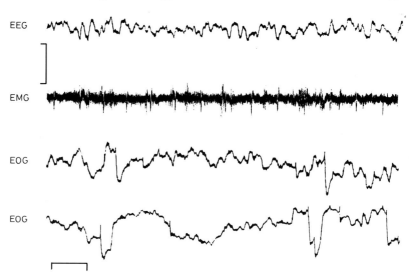

Fig. 3. Saccade spikes in connection with the rapid eye movements of the REM-stage of sleep. EOG is recorded with two derivations to detect conjugate eye movements. EMG is recorded from submental electrodes [*Rechtschaffen and Kales,* 1968].

The shape of the saccade spike resembles the M-response recorded in motor nerve conduction velocity measurements [*Goodgold and Eberstein,* 1978]. The shape and polarity of the spike could be caused by the simultaneous activation of several motor units in the orbicularis oculi muscle.

As shown in the results the amplitude and the shape of the saccade spike depends on the location of electrodes. In general, it is larger at the inner canthus than at the outer canthus during abducting saccades [*Jäntti,* 1982]. The different appearance of the saccade spike depending on the position of electrodes might explain some of the variation and asymmetries in the saccade velocities reported.

In the present study we demonstrate that even a prominent spike can be reduced by placing the electrodes just beside the outer canthi and by using low-pass filters. Even then, however, it is not totally eliminated, and, on the other hand, this position is impractical in clinical work. The possible existence of a prominent saccade spike should be taken into consideration especially when saccades are analyzed by computer.

References

Beaussart, M.; Guieu, J.D.: Artefacts; in Rémond, Handbook of electro-encephalography and clinical neurophysiology, vol. 11A (11A-80-11 A-87) (Elsevier, Amsterdam 1977).

Blinn, K.A.: Focal anterior temporal spikes from external rectus muscle. Electroenceph. clin. Neurophysiol. *7:* 299–302 (1955).

Goodgold, J.; Eberstein, A.: Electrodiagnosis of neuromuscular diseases; 2nd ed. (Williams & Wilkins, Baltimore 1978).

Jäntti, V.: Spike artefact associated with fast eye movements in electronystagmography and its importance in the automatic analysis of saccades. ORL (in press, 1982).

Rechtschaffen, A.; Kales, A.: A manual of standardized terminology, techniques and scoring system for sleep stages of human subjects. (Bethesda 1968).

Schmidt, D.; Thoden, U.: Co-activation of the M. transversus auris with eye movements (Wilson's oculo-auricular phenomenon) and with activity in other cranial nerves. Albrecht v. Graefes Arch. klin. exp. Ophthal. *206:* 227–236 (1978).

V. Jäntti, MD, Department of Pharmacology, Institute of Biomedicine,
University of Turku, Kiinamyllynkatu 10, SF-20520 Turku 52 (Finland)

Adv. Oto-Rhino-Laryng., vol. 30, pp. 76–79 (Karger, Basel 1983)

Electrooculography of Vertical Saccades

H.J. Scholtz, I. Pyykkö, N.G. Henriksson

ENT Clinic, Wilhelm Pieck University, Rostock, GDR, and ENT Department, University of Lund, Sweden

The investigation of horizontal saccades has proved its value in the diagnostics of various diseases of the brain stem. Electrooculography (EOG) proved to be a reliable method which yielded reproducible results. Principally, a similar importance can be attributed to testing voluntary vertical saccades provided that a suitable recording technique can be used. Our investigations focused on the question whether, due to the progress made in EOG, this method is still suitable for recording vertical eye movements. The EOG technique described by *Henriksson* et al. [1982] was used. A photoelectric nystagmography technique according to *Gestewitz* [1975] served as comparison in some investigations.

(1) Linearity in different gaze sectors at horizontal head position: Ten voluntary saccades of 20° each were performed, at horizontal head position, by 20 healthy volunteers in the upper, middle and lower gaze sector (fig. 1). The signal recorded corresponds to the middle sector of the calibration of 20°. In the upper sector, however, the signal is 23.3° higher, whereas an amplitude of 18° can only be recorded in the lower sector. The differences are highly significant from a statistical point of view.

(2) Linearity in different gaze sectors with changing head position: The saccade test of 20° in the three different sectors was carried out with raised or lowered head, respectively, in such a way that the outer targets could still be recognized well. The highest amplitudes of the signal could be observed with lowered head in the upper sector (fig. 2).

(3) Linearity with different saccade sizes: If voluntary vertical saccades are performed at an amplitude rising by 10°, there are differences to be observed depending on whether work is started at the high-

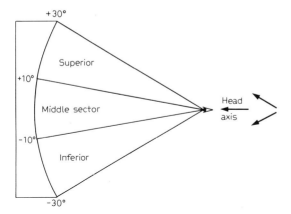

Fig. 1. Gaze sectors at horizontal and changing head position.

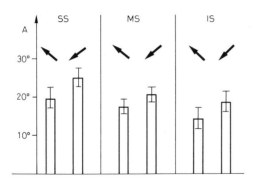

Fig. 2. Amplitudes (A) of signals in vertical saccadic test (saccades of 20°) with head position up (↖) or down (↙). SS = Superior sector; MS = middle sector; IS = inferior sector.

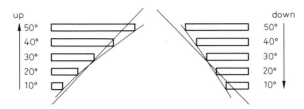

Fig. 3. Amplitudes of vertical saccadic signals when starting at the highest and lowest point of view.

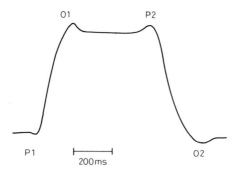

Fig. 4. EOG recording of 40° voluntary vertical saccades.

est or lowest point of gaze. The signal gain can be found higher if the amplitude increases from the lower to the upper position, whereas the signal gain will be smaller if the amplitude increases downwards (fig. 3).

(4) Artefacts: Figure 4 shows a typical EOG recording sample made at a vertical saccade of 40°. After a short opposite signal (P1), also to be found at voluntary horizontal saccade but shorter as described by *Aantaa and Jänti* [1982], the increase of the potential is frequently followed by an overshooting phenomenon (01). Before the decrease of the potential, which has to be attributed to the gaze movement, and increase of the potential (P2) is placed, whereas the following overshooting phenomenon (02) will occur less frequently, showing a smaller amplitude too. Table I informs about the frequencies and the average amplitudes of the potentials.

With photoelectric nystagmograms there are only overshooting phenomena, even in vertical saccades of 10°. A high fluctuation of potential can be deviated in EOG if the gaze is constantly concentrated on the target, with the eyelids being closed to a small opening. By means of a slight pressure on the deviating electrodes it could be ensured that there is no deviation at the muscle action. On opening the lid a fluctuation of the potential in the opposite direction can be observed.

(5) The peak signal velocities were greater at the upward gaze than at the downward gaze. At EOG recording the differences for 5, 10 and 20° saccades are statistically not significant, for 30, 40 and 60°, however, they are markedly significant. This is also true for those peak velocities determined by photoelectric nystagmograms of 10° saccades.

Table I. Frequencies (f) and amplitudes (A) of P1, O1, P2 and O2 potentials in voluntary vertical saccades (see text for explanation)

Saccades degrees	P1			O1			P2			O2		
	f %	A, degrees		f %	A, degrees		f %	A, degrees		f %	A, degrees	
		mean	SD		mean	SD		mean	SD		mean	SD
5	–			–			–			–		
10	–			–			–					
20	10	3.3	1.7	40	3.75	2.14	14	3.0	1.91	4	2.3	1.8
30	5	2.5	0.7	33	4.76	2.72	13	2.8	2.12	5	3.1	2.0
40	4	3.5	2.3	35	5.52	2.94	22	7.9	5.3	4	3.5	2.12
60	5	3.7	2.3	26	5.87	3.22	14	7.0	5.94	6	4.0	2.1

Conclusions

All findings obtained indicate that the conductance between eye electrodes is changed depending on the sector of vertical gaze. This may be due to interference of two factors; the position of the eye in relation to the boney wall is changed or the impedance due to eyelid position. In connective tissue the current for EOG spread more than in boney or fat tissue. The former has 20 times and the latter 2 times higher impedance than connective tissue.

Furthermore, the change of the form of the eyelids may have an influence in different ways as shown by *Berry and Melvill Jones* [1965]. The clinical application of the saccadic test with the recording technique used by us has to allow for the limitation that voluntary vertical movements of gaze should only be carried out to a maximum of 20°.

References

Berry, W.; Melvill Jones, G.: Influence of eye-lid movement upon elektro-oculographic recording of vertical eye movements. Aerospace Med. *36:* 855 (1965).

Henriksson, N.G.; Pyykkö, I.; Schalen, L.; Wennmo, C.: Velocity patterns of rapid eye movements. Acta oto-lar. *89:* 504 (1980).

H.J. Scholtz, MD, ENT Clinic, Wilhelm Pieck University, DDR-25 Rostock (GDR)

Adv. Oto-Rhino-Laryng., vol. 30, pp. 80–82 (Karger, Basel 1983)

Saccade and Pursuit Eye Movements

Setsuko Takemori [a], *Tadashi Aiba* [b], *Ryoichi Shiozawa* [c]

Departments of [a] Neurotology, [b] Neurosurgery, and [c] Neurology, Toranomon Hospital, Tokyo, Japan

Eye movements consist of two factors, namely saccades and pursuit eye movements. Saccades consist of quick movements of the eyes and are equal to the quick phase of nystagmus. Pursuit eye movements, on the other hand, consist of slow movements of the eyes and are equal to the slow phase of nystagmus. Saccades and pursuit eye movements are closely related to each other.

The purpose of this paper is to clarify which part of the central nervous system is essential for saccades and pursuit eye movements, respectively, especially in horizontal eye movements.

Methods

Eye movements were recorded by electrooculography using silver plate electrodes. To evoke saccades and pursuit eye movements simultaneously, the following stimuli were used; a round target, which was 1 cm in diameter, was moved from the right to the left side with a visual angle of 30° and a duration of 3 s. The target was moved back to the right at the moment when it reached the left side. The reversed movements of the target were also evoked.

Eye movements, saccades and pursuit eye movements were evoked by following the target movements.

Results

20 normal adults took part in the study. The speed of saccades was 169 ± 19°/s, and the latency between the initiation of the target movements and the eye movements was 0.15 ± 0.05 s.

Small and saccadic eye movements were 0–1 beat/cycle in normal subjects. Saccadic eye movements of more than 2 beats/cycle were seen in pathological cases.

The 256 pathological cases participating in the study were as follows:

Inner ear and vestibular nerve lesions (64 cases): In cases of Ménière's disease, sudden deafness, vestibular neuronitis, etc., normal saccades and normal pursuit eye movements were seen when spontaneous nystagmus was not seen. When spontaneous nystagmus to the contralateral side was seen, the saccades to the ipsilateral side and the pursuit eye movements from the ipsilateral side to the contralateral side were impaired.

Cerebellar lesions (44 cases): The speed of the saccades was normal in 95% of the cases of cerebellar lesions. The latency between the initiation of the target movements and the eye movements was prolonged in 60% of these patients. The pursuit eye movements were strongly impaired in all cerebellar lesions.

Brain stem lesions (32 cases): In cases of paramedian pontine reticular formation lesions, the saccades to the lesion side were strongly impaired while those to the contralateral side were slightly impaired. However, the pursuit eye movements to both sides were almost normal.

In cases of MLF syndrome, the saccades to the adduction of the lesion side eye were impaired. The saccades to the other directions and pursuit eye movements were intact. In cases of the other brain stem lesions, both saccades and pursuit eye movements were impaired.

Cerebellopontine angle lesions (47 cases): In cases of a small acoustic neurinoma of about 1 cm, the saccades and pursuit eye movements were almost normal. However, the impairment of the saccades and pursuit eye movements depended on the size of the tumor or the extension of the lesions.

Cerebral lesions (49 cases): The saccades were intact. However, the pursuit eye movements towards the ipsilateral side were impaired in frontal, occipital and upper parietal lobe lesions. In lower parietal lobe lesions, both the saccades and pursuit eye movements were impaired. In unilateral lower parietal lobe lesions, the saccades to the contralateral side were more strongly impaired than those to the ipsilateral side. The pursuit eye movements to the ipsilateral side were more strongly impaired than those to the contralateral side.

Finally, there were also 20 other cases.

Conclusion

Saccades. In unilateral paramedian pontine reticular formation lesions, the saccades to the ipsilateral side are completely impaired, while those to the contralateral side are slightly impaired. The pursuit eye movements are intact.

Pursuit Eye Movements. The pursuit eye movements from the ipsilateral side to the contralateral side are impaired in cases of unilateral inner ear or vestibular nerve lesions. Saccades and pursuit eye movements are closely related to each other. Therefore, the saccades to the ipsilateral side are sometimes impaired with the impairment of the pursuit eye movements from the ipsilateral side to the contralateral side when the spontaneous nystagmus to the contralateral side is very strong.

The pursuit eye movements are strongly impaired in cases of cerebellar lesions. The pursuit eye movements from the ipsilateral side to the contralateral side are impaired in cases of unilateral frontal, occipital and upper parietal lobe lesions.

Saccades and Pursuit Eye Movements. Both saccades and pursuit eye movements are impaired in cases of unilateral lower parietal lobe lesions and brain stem lesions except PPRF (paramedian pontine reticular formation) or MLF (medial longitudinal fasciculus) lesions.

S. Takemori, MD, Department of Neurotology, Toranomon Hospital,
Tokyo 105 (Japan)

Adv. Oto-Rhino-Laryng., vol. 30, pp. 83–87 (Karger, Basel 1983)

Examination of Optokinetic Nystagmus in Relation to Target Movement

Takashi Tokita, Tomoo Suzuki, Noboru Hishida, Masami Yanagida

Department of Otorhinolaryngology, Gifu University School of Medicine, Gifu, Japan

This report presents a method to examine the relationship between optokinetic nystagmus and target movement and the results obtained from the examination using the method in patients with spinocerebellar atrophy and congenital nystagmus.

Methods

An Ohm-type rotating cylinder was used to examine horizontal optokinetic nystagmus (OKN). The cylinder measured 2 m in diameter and 2 m in height. 12 vertical stripes were drawn at equal intervals on its inner surface. The cylinder was rotated electrically with an angular acceleration of $2°/s^2$ for 90 s.

OKN, induced by the optokinetic stimulation, and signals, indicating that stripes crossed in front of the subject, were simultaneously recorded by an electronystagmograph and a magnetic data recorder. The data on the tape were sampled at 100 Hz for 90 s by a PDP-12 computer through an analog-to-digital converter. The stripe movements were calculated from stripe signals using Lagrange's square interpolation formula. The nystagmus waves and stripe movement were displayed in superimposition on a cathode ray tube. From the relation between the two, the presence of inverted nystagmus and the ocular ability to catch and follow the stripes were evaluated and printed on a teletypewriter. The catching ability was classified into fit, overshoot or undershoot. The pursuit ability, on the other hand, was classified into corresponding, delay or preceding types. In addition, the eye pursuit itself was evaluated as smooth, saccadic or irregular.

In normal subjects, at a cylinder speed of up to 40°/s, pursuit eye movement corresponded to target movement and target catch was done exactly. Moreover, up to a cylinder speed of 100°/s, nystagmus appears at every stripe. OKN in patients were evaluated in comparison with those of normal subjects.

Results

OKN of Patients with Spinocerebellar Atrophy
The upper frames of figure 1 indicate the OKN of a patient with late cerebellar cortex atrophy (LCCA). At a cylinder speed of 0–10°/s, eye pursuit is characterized by a square wave jerk. From 10 to 31°/s, a large amplitude saccadic movement (macrosaccade) in stripe pursuit, and an overshoot in target catch are noted. At a cylinder speed of 31–41°/s, stepped saccade due to ocular dysmetria is observed in the eye movement for the target catch, that is the rapid phase of nystagmus. Moreover, when the cylinder speed is increased, rapid phase generation is insufficient and nystagmus occurs sporadically.

The lower frames represent the OKN of a patient with olivopontocerebellar atrophy (OPCA). At a cylinder speed of 0–20°/s, the stripe movement is followed by eye movement composed of repetition of a small amplitude saccadic movement (microsaccade). At a cylinder speed of 21–31°/s, delayed initiation of target catching and reduced saccadic velocity are observed. At a cylinder speed of 31–41°/s, the pursuit eye movement is clearly behind the stripe movement. However, target fixation by the retinal fovea is performed accurately, and it is difficult to consider ocular dysmetria present. When cylinder speed is further increased, pursuit of stripe movement becomes virtually impossible, and the number of nystagmus beats markedly decreases.

OKN abnormalities in 7 cases diagnosed as LCCA and 6 diagnosed as OPCA were as follows. In stripe pursuit, the delay of eye movement and deficit of smooth pursuit are observed with a similar frequency in both groups. However, there is a tendency for LCCA patients to indicate macrosaccade and for OPCA patients show microsaccade. In target catching, ocular dysmetria is frequently observed in both groups. However, in 3 cases of OPCA, even when smooth pursuit is hindered, signs of ocular dysmetria are not observed. Delayed initiation of target catching and reduced saccadic velocity were only observed in 2 cases of OPCA.

The difference between OKN abnormality characteristics in cases with LCCA and OPCA is presumably that, in LCCA, degeneration in Purkinje cells of the cerebellar vermis is predominant, whereas in OPCA there is also degeneration of the pontine nuclei.

OKN abnormalities observed in cases with spinocerebellar atrophy were reported by *Jung and Kornhuber* [1964], *Alpert* et al. [1975],

Late cerebellar cortex atrophy, 52-year-old male

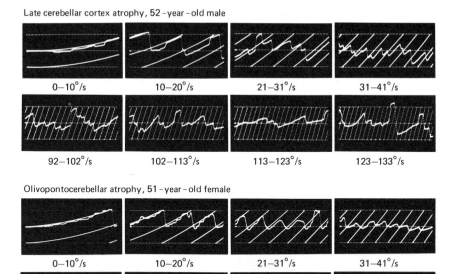

| 0–10°/s | 10–20°/s | 21–31°/s | 31–41°/s |

| 92–102°/s | 102–113°/s | 113–123°/s | 123–133°/s |

Olivopontocerebellar atrophy, 51-year-old female

| 0–10°/s | 10–20°/s | 21–31°/s | 31–41°/s |

| 41–51°/s | 51–61°/s | 61–72°/s | 72–82°/s |

Fig. 1. OKN abnormalities in patients with spinocerebellar atrophy. OKN in a patient with LCCA shows square wave jerks, saccadic eye movements in stripe pursuit and stepped saccade due to ocular dysmetria in the eye movement for the target catch. OKN in a patient with OPCA shows irregular pursuit, delayed initiation of target catching and reduced saccadic velocity.

Baloh et al. [1975], *Zee* et al. [1976] and *Murphy and Goldblatt* [1977]. These studies have reported normal or reduced OKN in patients with cerebellar cortex atrophy and disturbances of saccadic movement generation in OPCA patients. We consider that the examination of the relationship between OKN and target movement demonstrated in detail the peculiarity of OKN abnormalities in cases with LCCA and OPCA.

Inverted OKN of Patients with Congenital Nystagmus

Figure 2 shows the inverted OKN encountered in cases with congenital nystagmus. In the upper frames of the figure, at a cylinder speed of 0–20°/s, the eye follows the stripe movement with repetition of a saccadic movement characterized by a small amplitude. At a cylinder speed of 21–31°/s, the moving stripe is followed by a large ampli-

Inverted OKN, 9 - year - old male

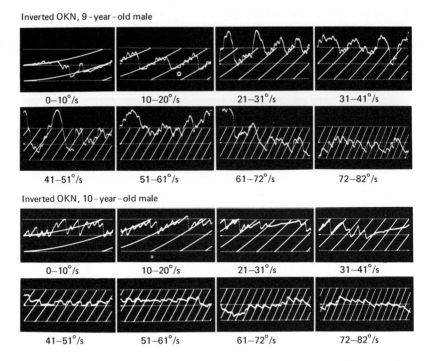

Fig. 2. Inverted OKN in patient with congenital nystagmus. Upper frames: Stripe pursuit with saccadic eye movements develops into inverted OKN with increase of stripe velocity. Lower frames: At a cylinder speed of 61–82/s, the eye produces a saccadic eye movement in the direction of stripe movement when the fovea fits a stripe. The saccadic pursuit forms the inverted OKN.

tude saccadic eye movement. As the cylinder speed increases from 31 to 72°/s, the stripe pursuit is performed by various saccadic eye movements of different amplitudes. When the cylinder speed surpasses 72°/s, the amplitude and rhythm become uniform so as to develop into typical inverted OKN. This finding suggests that the inverted OKN is due to stripe pursuit with saccadic eye movement.

The lower frames indicate the inverted OKN observed in a case showing active spontaneous nystagmus. At a cylinder speed of 0–20°/s, pursuit movement superimposed on the spontaneous nystagmus is observed. The pursuit movement seems to be a pursuit produced by the action of the peripheral retina. At a cylinder speed of 61–82°/s, the eye produces a saccadic eye movement in the direction of stripe movement when the fovea fits a stripe. Then, the moving stripe is followed by saccadic movement instead of smooth pursuit. The saccadic eye movement forms the inversed OKN.

Based on the reports by *Barany, Brunner* and *Borries, Duke-Elder and Wybar* [1973] mentioned that the inversion is due to the fact that the pre-existing nystagmus takes precedence over the optokinetic phenomenon and may thus augment it or interfere with it. *Hood and Leech* [1974] reported that the dependence of the slow component of the nystagmus upon peripheral stimulation may be invoked by way of explanation of the phenomenon of reversed OKN. *Yee* et al. [1980] suggest that there is a defect in the subcortical optokinetic system in congenital nystagmus.

From the foregoing observation, we considered that, because of the foveal inability to perform smooth pursuit eye movement, the inverted OKN resulted from the target pursuit with saccadic eye movement. Moreover, on the decrease of the fast phase velocity, we supposed that it depends on such a peculiarity as the eye which moves rapidly in one direction is hard to move rapidly, without interruption, in the opposite direction.

References

Alpert, J.N.; Coats, A.C.; Perusquina, E.: Saccadic nystagmus in cerebellar cortical atrophy. Neurology, Minneap. *25:* 676–680 (1975).

Baloh, R.W.; Konrad, H.R.; Honrubia, V.: Vestibulo-ocular function in patients with cerebellar atrophy. Neurology, Minneap. *25:* 160–168 (1975).

Duke-Elder, S.; Wybar, K.: Optokinetic nystagmus. System of ophthalmology, pp. 808–812 (Kimpton, London 1973).

Hood, J.D.; Leech, J.: The significance of peripheral vision in the perception of movement. Acta oto-lar. *77:* 72–79 (1974).

Jung, R.; Kornhuber, H.: Results of electronystagmography in man; in Bender, The oculomotor system, pp. 440–442 (Harper & Row, New York 1964).

Murphy, M.J.; Goldblatt, D.: Slow eye movements, with absent saccades, in a patient with hereditary ataxia. Archs Neurol., Chicago *34:* 191–195 (1977).

Yee, R.D.; Baloh, R.W.; Honrubia, V.: Study of congenital nystagmus: optokinetic nystagmus. Br. J. Ophthal. *64:* 926–932 (1980).

Zee, D.S.; Yee, R.D.; Cogan, D.G.; Robinson, D.A.; Engel, W.K.: Oculomotor abnormalities in hereditary cerebellar ataxia. Brain. *99:* 207–234 (1976).

T. Tokita, MD, Department of Otorhinolaryngology,
Gifu University School of Medicine, Gifu 500 (Japan)

Adv. Oto-Rhino-Laryng., vol. 30, pp. 88–93 (Karger, Basel 1983)

Decrease of Integration of Optokinetic Nystagmus of Peripheral Retinal Type after Hemilabyrinthectomy

I. Pyykkö[a], *I. Matsuoka*[b], *S. Ito*[b], *M. Hinoki*[b]

[a] Department of Otorhinolaryngology, University Hospital of Lund, Sweden, and
[b] Department of Otorhinolaryngology, University of Kyoto, Japan

Introduction

Depending on the phylogenetic stage, different optokinetic mechanisms are present. In non-foveate animals, e.g. in rabbits, a subcortical optokinetic nystagmus (OKN) has been found [*Collewijn,* 1971]. In primates, which have well-developed fovea, a cortical and a subcortical OKN have been identified [*Dichgans,* 1977]. The subcortical OKN seems to utilize accessory optic pathways and is bound on activation of peripheral retinal receptors. Accordingly, it is also called the 'stare' type of OKN. The cortical OKN seems to utilize geniculo-striate pathways and hence the smooth pursuit mechanisms. Accordingly, the cortical OKN is also called the 'foveal' type of OKN or the 'look' type of OKN.

The effect of alertness on OKN has been a subject of controversy. Based on experiments in rabbits, *Collewijn* [1971], among others, concluded that attention does not play any significant role in the subcortical type of OKN. However, an increase of attention in non-foveate animals could increase their optokinetic gain [*Ter Braak,* 1936; *Pyykkö* et al., 1982b, c]. In the experiments of the latter authors, the increased optokinetic responses could be related to activation of EEG. Especially vibratory stimulus was powerful in evoking high velocity OKN. Concomitant with an increase in velocity of OKN, an activation of EEG in the form of theta waves was observed.

A response decline in OKN has been observed after labyrinthectomy [*Collewijn,* 1976], which may, at least partly, be the result of decreased vigilance due to lack of activation of the brain stem by vestibular impulses [*Pyykkö* et al., 1982b]. In the present work, the effect of attention and unilateral labyrinthine lesion on the OKN and the interaction of optokinetic and vestibular (optovestibular) nystagmus were examined in rabbits.

Material and Methods

Data were collected from 10 intact and 6 hemilabyrinthectomized black or brown Japanese rabbits. Detailed description of the methods for implanting of permanent EEG electrodes, hemilabyrinthectomy and tests have been published elsewhere [*Pyykkö* et al., 1982a–c].

The tests were conducted 8–12 weeks after hemilabyrinthectomy. During the tests the rabbit was placed in a plexi cabin with the head immobilized with a collar and mouth piece. The cabin was rotated at a constant velocity of 60°/s either to the right or the left. The full-field optokinetic drum was driven separately at a constant velocity of 60°/s. Rotation was also performed in light at a constant velocity of 60°/s, the optokinetic drum as a stationary visual field. As alerting stimuli local vibration to the abdomen was applied with a rotatory type of vibrator at frequency range of 20–140 Hz and at a constant amplitude of 0.9 mm.

Results

Optokinetic Test

In intact rabbits the OKN was considerably weak. Most commonly OKN started with the fast component. By applying vibration a highly significant increase in the velocity ($p < 0.001$) and frequency ($p < 0.001$) of the nystagmus was observed in all rabbits (fig. 1). The increase was linked to activation of the EEG observed as theta waves. Even though during exposure to vibration the velocity of OKN powerfully increased, the rise time and the maximum response in OKN were different from those of vestibular and optovestibular nystagmus (fig. 2).

In the hemilabyrinthectomized rabbits the frequency of OKN was significantly lower than in intact animals ($p < 0.01$; fig. 3). Decrease in the frequency was present in both nystagmus directions. Exposure to vibration only marginally increased the frequency of OKN in hemilabyrinthectomized rabbits. During the test-free period sudden exposure to vibration was able to cause nystagmus in most labyrinthectomized animals, which was only occasionally observed in intact rabbits.

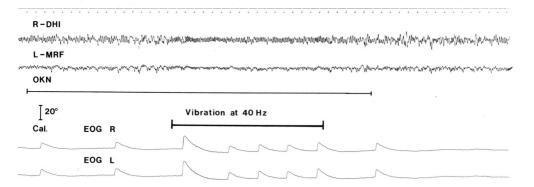

Fig. 1. Activation of EEG and eye movements during vibratory stimulus. Upper-
most tracing indicates time, one division is 1 s. Middle tracings: EEG recording from
right dorsal hippocampus (R-DHI) and left midbrain reticular formation (L-MRF).
Lowest tracings: eye movement recording from the right (EOG R) and the left (EOG L)
eye.

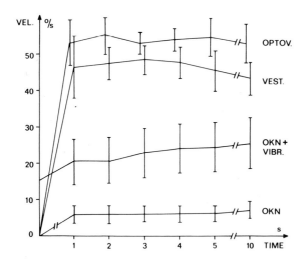

Fig. 2. Mean velocity of eye movements during exposure to optokinetic (OKN) and
optokinetic and vibratory (OKN + vibr.) stimuli, angular acceleration in the darkness
(vest.) and angular acceleration in light within the stationary optokinetic drum (optov.).
Mean velocity and standard deviation of eye movements for each second is exhibited.
For OKN the first observed beat is exhibited in the first second.

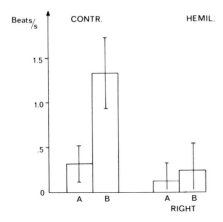

Fig. 3. Mean maximum frequency of nystagmus during optokinetic test without vibration (A) and with vibration (B). In hemilabyrinthectomized rabbits only OKN to the right is exhibited.

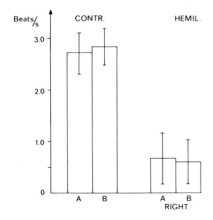

Fig. 4. Mean maximum frequency of optovestibular nystagmus in intact and hemilabyrinthectomized rabbits. Abbreviations as in figure 3.

Optovestibular Test

Rotating an intact rabbit in light within the optokinetic drum caused a regular and non-fatiguable nystagmus. The frequency (3.0 beats/s) was slightly higher than the frequency of vestibular nystagmus (2.8 beats/s). Exposure to vibration only slightly increased the frequency of nystagmus (3.2 beats/s; $p < 0.05$), but the velocity of nystagmus remained the same.

Optovestibular responses in hemilabyrinthectomized rabbits were poor and weaker than in intact rabbits in both directions ($p < 0.01$; fig. 4). There was also a directional preponderance in the responses. The nystagmus towards the hemilabyrinthectomized side was weaker than the nystagmus towards the intact side ($p < 0.01$). Noteworthy is that exposure to vibration in hemilabyrinthectomized rabbits could not enhance optovestibular responses.

Discussion

In an intact animal attention is an important determinant of opto-kinetic output. As well known, vestibular output is also dependent on attention. Interestingly, in the present experiments the activation of the vestibular system by rotation in the darkness lead to an immediate increase of EEG activity in rabbits. Thus, higher attention increases vestibular responses and vestibular stimulation leads to increase of attention. Attention and vestibular mechanisms are interlocked.

The importance of the labyrinth for vigilance of the animal seems to be evident from the animal behavior after destroying the labyrinth. In the present study the rabbits refused to drink or eat for several days after operation and more than half of the operated rabbits died. According to *Collewijn* [1976] mortality is 100% after bilateral labyrinthectomy. It seems very likely that the tonic output of the vestibular organs is one important part of the homeostatic system of the animal.

The similarity of disorders in vestibular nystagmus and in OKN of non-foveate animals suggests that the optokinetic and the vestibular mechanisms share much the same neuronal circuitry. The brain stem mechanisms integrating these responses seem not to be identical but complementary to each other as suggested by the difference in maximum velocity and the time constant of vestibular and optokinetic responses. In optovestibular nystagmus the labyrinthine stimulation seems to provide activation of brain stem mechanisms to ensue proper optokinetic gain and thus stabilization of visual surrounding during active head movements.

After a labyrinthine lesion a response decline occurs bilaterally in the neurons of the vestibular nuclei [*McCabe* et al., 1972]. The response decline is modified by a regeneration of the neural activity in the deafferented side and by a release of the inhibition in the intact side. Thus, after cessation of the spontaneous nystagmus the vestibular responses become more symmetric, although strongly reduced. The brain mechanisms controlling the compensation of neural activity seem to operate through the cerebellum. An analogous or identical mechanism as in the vestibular nystagmus seems to operate in OKN of the peripheral retinal type as indicated in present results. A similar response decline was also observed in optovestibular nystagmus. In the hierarchy of nystagmic responses, the optokinetic part seems to be under the vestibular part. Moreover, optovestibular nystagmus seems to be composed of a mere algebraic sum of vestibular and optokinetic nystagmus both in intact and hemilabyrinthectomized rabbits.

References

Collewijn, H.: The optokinetic system of the rabbit. Documenta ophth. *30:* 205–226 (1971).

Collewijn, H.: Impairment of optokinetic (after-) nystagmus by labyrinthectomy of rabbit. Expl Neurol. *52:* 146–156 (1976).

Dichgans, J.: Optokinetic nystagmus as dependent on the retinal periphery via the vestibular nuclei; in Baker, Berthoz, Control of gaze and brain stem neurons: developments in neuroscience, vol. 1, pp. 261–267 (Elsevier/North-Holland Biomedical Press, Amsterdam 1977).

McCabe, B.F.; Ryu, J.H.; Sekitani, T.: Further experiments of vestibular compensation. Laryngoscope, St Louis *82:* 381–396 (1972).

Pyykkö, I.; Magnusson, M.; Matsuoka, I.; Ito, S.; Hinoki, M.: Effects of pure-tone sound, impulse noise and vibration on visual orientation. Am. J. Otolar. *3:* 104–111 (1982a).

Pyykkö, I.; Magnusson, M.; Matsuoka, I.; Ito, S.; Hinoki, M.: On the optokinetic mechanism of peripheral retinal type. Acta oto-lar., suppl. 386, pp. 235–239 (1982b).

Pyykkö, I.; Matsuoka, I.; Ito, S.; Hinoki, M.: Enhancement of eye motor response and electrical brain activity during noise and vibration in rabbits. Otolaryngol. Head Neck Surg. *90:* 130–138 (1982c).

Ter Braak, J.W.G.: Untersuchungen über optokinetischen Nystagmus. Archs neerl. Physiol. *21:* 309–376 (1936).

I. Pyykkö, MD, Department of Otorhinolaryngology, University Hospital of Lund, S-221 85 Lund (Sweden)

Adv. Oto-Rhino-Laryng., vol. 30, pp. 94–97 (Karger, Basel 1983)

The Effect of Central Retinal Lesions on Nystagmus in the Monkey during Visual-Vestibular Conflict Stimulation[1]

U. Büttner[a], *O. Meienberg*[b], *B. Schimmelpfennig*[c]

[a] Department of Neurology, University of Düsseldorf, FRG;
[b] Department of Neurology[2], University of Bern, Switzerland, and
[c] Department of Ophthalmology, University of Zürich, Switzerland

A visual-vestibular conflict situation can be achieved experimentally by coupling a turntable mechanically to its surrounding visual environment, for example a striped cylinder. A monkey seated on this turntable experiences a sensory conflict during rotation since vestibular stimulation occurs without displacement of the visual surroundings. This visual-vestibular conflict leads to a marked suppression of the vestibulo-ocular response (VOR-Supp) [6]. It is thought that the same neuronal mechanisms are used for both optokinetic nystagmus (OKN) and this VOR-Supp. It has recently been shown, quantitatively, that high-velocity OKN consists of two components [3]: a 'fast' component, which is responsible for an initial rapid rise in eye velocity, and is thought to utilize the smooth pursuit system [4] and second, the slow velocity build up during continuous OKN-stimulation, which is mediated by the 'velocity storage' component [3].

Recent single unit studies in alert monkeys are in accordance with these findings [2, 7]. From these studies it can also be concluded that the VOR-Supp during visual-vestibular conflict stimulation is mainly due to activation of the 'fast' component. Since the smooth pursuit system is involved in foveal tracking of small visual objects, dysfunction of the fovea should severely affect VOR-Supp. We found that foveal lesions had little effect.

[1] This work was supported by Swiss National Foundation 3.343-2.78 and Deutsche Forschungsgemeinschaft SFB 200, A 2.
[2] The experiments for this work were performed at the Department of Neurology, University of Zürich.

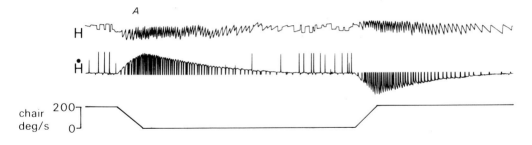

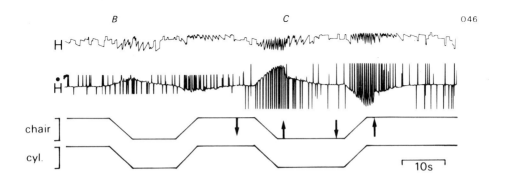

Fig. 1. Eye movement recordings of a monkey during vestibular stimulation in the dark and visual-vestibular conflict stimulation *(B, C)*. The right eye (with a lesion of 10–12° centered at the fovea) was exposed, the left eye was covered by a patch. Traces show from above: horizontal eye position, upward deflection indicates movement to the right; eye velocity (first derivative of eye position) where velocity of fast phases is clipped-off at an arbitrary level. Bottom: stimulus velocity. For stimulation the monkey is rotated around a vertical axis. *A* Acceleration and deceleration in the dark lead to nystagmus which slowly and monotonically decreases after the end of stimulation. *B, C* A cylinder is mechanically coupled to the turntable. *B* The monkey is rotated in the light which leads to a marked decrease of the nystagmus (compare with *A*). *C* Downward arrows indicate 'light off' and upward arrows 'light on'. Thus, during vestibular stimulation in the dark nystagmus velocity builds up, but then shows a rapid decrease after 'light on'. Almost identical responses are found for the normal eye, indicating that central retinal lesions have very little effect on VOR-Supp.

Experiments were performed on alert monkeys *(Macaca mulatta)*. During the experiments the monkey sat, with its head stabilized, on a turntable, which was completely surrounded by a cylinder covered with vertical black and white stripes (width 7.5°). During stimulation, turntable and cylinder were mechanically coupled and rotated around the vertical axis with velocity trapezoids (acceleration $40°/s^2$, constant velocity $200°/s$).

Confluent laser lesions between 6 and 12° diameter, centered at the fovea, were placed in one eye. In the monkey, fovea and parafovea have a diameter of 6° [5]. The responses of the normal and lesioned eye were then compared.

Two conditions of VOR-Supp (VOR-Supp I and VOR-Supp II) were investigated. During VOR-Supp I the whole stimulus sequence occurred in the light: as figure 1B shows, this leads to a pronounced decrease in nystagmus velocity and duration even though the record was taken during stimulation of the lesioned eye (lesion diameter 10–12°). In the VOR-Supp II paradigm, the monkey is accelerated in the dark before the light inside the cylinder is switched on. Facing the motionless (relative to the monkey) cylinder, this leads to a sharp decrease in eye velocity, which similarly can be seen after deceleration (fig. 1C). A comparison of the responses from the normal and lesioned eye showed that central retinal lesions had only minor effects on nystagmus suppression in both paradigms.

Comments

The results clearly demonstrate that in the monkey VOR-Supp in visual-vestibular conflict paradigms is not greatly affected by central retinal lesions up to 12° diameter. It has been pointed out earlier that all evidence suggests that the suppression of the VOR is the result of the activation of the 'fast' OKN component. This is further supported by the rapid change in eye velocity after light on seen in the VOR-Supp II experiment (fig. 1C), which could not be achieved so rapidly by the 'velocity storage' component. In addition, direct OKN measurements in the monkey show that the 'fast' component is hardly affected by central retinal lesions of this size [1]. Thus both the experiments on VOR-Supp and direct OKN measurements underline the fact that the visual input to the 'fast' OKN component also originates from retinal areas outside the fovea, areas generally assumed not to be involved in smooth pursuit eye movements. However, these findings do not necessarily conflict with the assumption of *Robinson* [4]. They rather suggest that also large areas outside the fovea contribute a visual input to the smooth pursuit system.

References

1   Büttner, U.; Meienberg, O.; Schimmelpfennig, B.: in Roucoux and Grommelinck, Physiological and pathological aspects of eye movements, pp. 173–179 (Junk, The Hague 1982).
2   Büttner, U.; Waespe, W.; Henn, V.: in Huber, Klein, Neurogenetics and neuro-ophthalmology, pp. 89–102 (Elsevier/North-Holland Biomedical Press, Amsterdam 1981).
3   Cohen, B.; Matsuo, V.; Raphan, T.: J. Physiol. *270:* 321–344 (1977).
4   Robinson, D.A.: in Henn, Cohen, Young, Visual-vestibular interaction in motion perception and the generation of nystagmus. Neurosciences Research Program Bull. 18, Nr. 4 (1980).
5   Stone, S.: J. comp. Neurol. *124:* 337–352 (1965).
6   Waespe, W.; Henn, V.: Exp. Brain Res. *33:* 203–211 (1978).
7   Waespe, W.; Henn, V.: Exp. Brain Res. *43:* 349–360 (1981).

U. Büttner, MD, Department of Neurology, University of Düsseldorf,
D–4000 Düsseldorf (FRG)

Adv. Oto-Rhino-Laryng., vol. 30, pp. 98–101 (Karger, Basel 1983)

Comparison between Narrow-Angle Esocentric and Wide-Angle Egocentric Optokinetic Nystagmus

S. Holm-Jensen, J. Thomsen

University ENT Department, Hvidovre Hospital, Hvidovre, Denmark

Optokinetic nystagmus (OKN) may be elicited by two different modes of stimulation: (1) by the optokinetic drum stimulating only the central vision, the subject is situated away from the center of rotation, and (2) by light slits projected onto a circular screen covering both the central and the peripheral vision, the subject is placed in the center of rotation.

The aim of this study was to compare the optokinetic responses to the two different ways of stimulating OKN in a group of patients with pronounced optokinetic defects.

Material and Method

50 patients with multiple sclerosis participated in the study (age 32–68 years). The duration of disease varied from 5 to 35 years. Binocular, horizontal eye deflections were recorded by ENG (a.c. amplification with a time constant of 2.5 s). Two parameters were used as criteria for pathology: (1) the rhythm, only a continuous, even response without pause and without distinct variation in the slope of the slow phase, was considered normal, and (2) the eye velocity of the slow phase (EV) averaged over a period of 10 s. Wide-angle OKN was elicited by a rotator which projected light slits onto a semicircular screen. Two target speeds were investigated, 10 and 20°/s. The normal OKN with this equipment has been published previously [*Holm-Jensen* et al., 1981]. Narrow-angle OKN was elicited by a drum with a diameter of 15 cm, placed 50 cm away from the patient. Three fixed speeds were used: 72, 180 and 360°/s, corresponding to a relative speed of the targets of 11, 27 and 54°/s, respectively. A requirement of less than 10% difference between clockwise and counterclockwise velocity of EV was found to be inadequate, as

the normal range increases with the target speed. The 95% prediction area with 72°/s was 6.1–10.6°/s, and with 180°/s, 7.9–17.3°/s. Of 20 normal subjects, 10 were unable to perform an even, continuous OKN at 360°/s, a normal range at this speed was unattainable. All had even responses at the other two speeds; 2 subjects were out-liers with regard to the symmetry of the responses at 72 and 180°/s, and an additional 4 subjects were out-liers at 360°/s.

Results

Wide-Angle Egocentric Stimulation. 38 patients had abnormally low EV in one or both directions or had dysrhythmic responses to tests with 20°/s. This speed proved to be a more difficult task than 10°/s, at which only 20 patients had abnormal OKN ($p < 0.01$). Furthermore, 7 patients with unilateral depletion at 10°/s had bilateral defects at 20°/s.

Narrow-Angle Esocentric Stimulation. 37 patients had abnormal optokinetic responses at 180°/s, and 29 of these had abnormal OKN also at 72°/s ($p < 0.05$). As was the case with wide-angle OKN, the slower speed did not reveal the full pathology. 7 patients with bilateral defective OKN at 180°/s had unilateral defects at 72°/s. The response at 360°/s was regular and symmetric in 8 patients, but owing to the large variations in the normal subjects, no conclusions could be drawn. Consequently, tests with 360°/s were rendered insignificant.

Concordance between Wide-Angle and Narrow-Angle OKN. The comparison between wide-angle and narrow-angle OKN was based on tests with 20°/s and 180°/s. Although quantitatively there was a high degree of concordance between results with the two methods, 10 patients had qualitative differences. 9 patients were unable to perform OKN at all. 4 patients had bilateral depletion at 20°/s, but unilateral depletion with 180°/s. 3 patients were listed as OKN normal with narrow-angle stimulation, but as pathological with wide-angle stimulation, and in all cases the EV was just above the normal range for narrow-angle OKN in one direction. This difference must be attributed to the large range of the normal response (type-II error). 1 patient had a unilateral defect with narrow-angle OKN and a bilateral defect with wide-angle OKN. There was no obvious explanation for this difference.

Interference by Spontaneous Nystagmus

Wide-Angle Stimulation. Spontaneous nystagmus (SN) was present (2.5–7.1°/s) in 4 of 12 patients with unilateral depletion either at 10 or 20°/s. In three cases it was ipsidirectional to the optokinetic preponderance and in one case beating in the opposite direction. Of the 12 patients with normal response, SN (2.7–7.3°/s) was present in five cases, and in three of these OKN was weaker in the direction of SN.

Narrow-Angle Stimulation. SN (2.5–7.3°/s) was present in 12 patients without abolished OKN at 180°/s. All cases had preponderance in the direction of the SN, in four cases as a unilateral depletion. Among these were two cases with normal, wide-angle OKN.

Discussion and Conclusion

An optokinetic defect is a frequent finding in multiple sclerosis. With an esocentric narrow-angle technique, *Noffsinger* et al. [1972] and *Dam* et al. [1975] reported optokinetic symptoms in 55–75% of severe cases. These figures are confirmed in the present study, but more detailed information is obtained with wide-angle stimulation. Discrepancies between results with the two of modes of stimulation could be explained by the wide range of the normal responses with narrow-angle stimulation, in addition to interaction from SN, probably owing to the lower gain. Wide-angle OKN was less influenced by interference from SN, which is in accordance with previous findings [*Holm-Jensen*, 1982]. The target speed is essential for the degree of information obtained from OKN. A slow speed will disclose the pathology in part only (type-II error), while a fast speed is too insensitive because of the wide range of normal responses (type-I error). Therefore, the three fixed speeds commonly used with narrow-angle OKN should be avoided.

References

Dam, M.; Johnsen, N.J.; Thomsen, J.; Zilstorff, K.: Vestibular aberrations in multiple sclerosis. Acta neurol. *52:* 407–416 (1975).

Holm-Jensen, S.: Interference between synchronous optokinetic nystagmus and vestibu-
  lar nystagmus. Acta oto-lar. *93:* 375–385 (1982).
Holm-Jensen, S.; Skovgaard, L.T.; Peitersen, D.: Synchronous optokinetic nystagmus.
  The prediction area in normal humans. Acta oto-lar. *91:* 255–266 (1981).
Noffsinger, D.; Olsen, W.O.; Carhart, R.; Hardt, C.; Sahgal, V.: Auditory and vestibular
  aberrations in multiple sclerosis. Acta oto-lar., suppl. 303 (1972).

S. Holm-Jensen, MD, University ENT Department, Hvidovre Hospital,
DK-2650 Hvidovre (Denmark)

Adv. Oto-Rhino-Laryng., vol. 30, pp. 102–107 (Karger, Basel 1983)

Optokinetic Stimulation and Body Balance

Toyoji Miyoshi, Akira Tamada

Department of Otolaryngology, Faculty of Medicine, Kyoto University, Kyoto, Japan

It is well known that optokinetic (OK) stimulation induces dizziness and influences body balance. When subjects look at a fixed point which has been set between the eye and moving surroundings, the dizzy sensation is much stronger than when there is no point of fixation [*Fukuda*, 1957]. *Miyoshi* et al. [1979] reported biphasic body sway by OK stimulation, and the direction of body deviation did not depend upon acceleration, but upon the velocity of OK stimulation. There are several differences between nystagmus induced by foveal and peripheral OK stimulation [*Miyoshi* et al., 1978]. In this paper, body sway induced by several OK stimulations is discussed.

Method

A force plate for measuring the center of gravity of the body is put just under the projector of OK stimulator. The separating device for visual field is set on the projector. The subjects stand on the force plate with their feet together and face the stimulator screen. The projector is driven at a uniform acceleration of $1°/s^2$. Four conditions of OK stimulation are used: (1) OK stimulation of the whole visual field; (2) peripheral OK stimulation in the looking condition; (3) peripheral OK stimulation in the staring condition, and (4) OK stimulation of the whole visual field with a fixation point. A red light emission diode (LED) is set on the screen of the OK stimulator. The body sway, direct current (DC) recorded OK nystagmus and photo marks which indicate the moment when the stripes pass through the middle of the screen are registered simultaneously on a DC nystagmograph. Target movements are drawn on the nystagmogram afterwards by means of an overlapping method [*Miyoshi* et al., 1978] with the help of the photo marks. Thus the relationship between targets and eye movements can be seen directly on the nystagmogram.

$1^{\circ}/s^2$, $0-120^{\circ}/s$

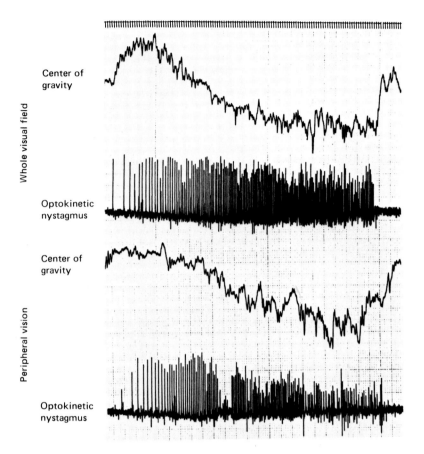

Fig. 1. Body sway by optokinetic stimulation (to the left).

Result

Usual OK Stimulation of Whole Visual Field

The biphasic body deviation can be seen (fig. 1). The body sway increases its amplitude and frequency according to the increase in stimulation velocity. In the slow–speed range of stimulation, the body deviates to the counter-direction of stimulation and then changes the direction of deviation and is attracted to the same direction of stimulation in

$1°/s^2, 0–120°/s$

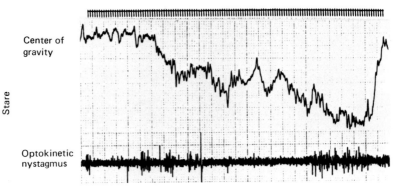

Stare

Center of
gravity

Optokinetic
nystagmus

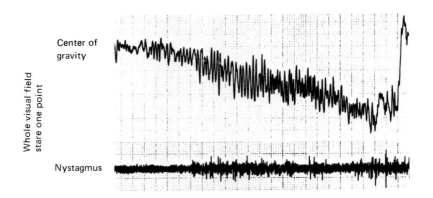

Whole visual field
stare one point

Center of
gravity

Nystagmus

Fig. 2. Body sway and optokinetic stimulation (to the left).

the high-speed range. In the first phase the amplitude and frequency
are not so big. In the second phase the amplitude and frequency increase remarkably. The OK nystagmus is well-induced (fig. 1).

Looking at the Peripheral OK Stimulation

Biphasic deviation cannot be seen any more. The body deviates to
the same direction of stimulation even in the slow-speed range. OK
nystagmus is also well-induced (fig. 1).

Staring in One Direction in Peripheral OK Stimulation
Biphasic body deviation cannot be seen in this test condition. The body is attracted to the same direction of stimulation from the beginning of rotation. No nystagmus is induced (fig. 2).

Fixation upon a LED on the Screen in OK Stimulation
In this test condition biphasic body deviation is never seen. The body balance is disturbed by OK stimulation even in the slow-speed range and attracted to the same direction. As fixation is complete, nystagmus cannot be induced (fig. 2).

Discussion

The movement of surroundings will induce the illusion of body motion in a counter-direction and may result in the bodily movement [*Fischer and Veits*, 1927; *Vogel*, 1931]. *Miyoshi* et al. [1979] investigated quantitatively the influence of OK stimulation upon body sway and concluded that the body sway induced by OK stimulation shows biphasic deviation according to the stimulating velocity. When the velocity is slower than 45°/s, the body deviated to the same direction of stimulation. The amplitude and frequency of body sway in the former phase are not so big. This fact indicates the body balance in this phase is well maintained and coordinated. The amplitude and frequency in the latter phase are, however, quite big. This fact indicates that the latter phase is a disturbed one. Many authors have reported that the upper limit of eye velocity to follow the moving target exactly is 40–60°/s [*Aschan* et al., 1955; *Miyoshi and Pfaltz*, 1973]. In the case of stimulation faster than this limit, the subject cannot keep the target image on his fovea and the image slips on the retina. For the correct orientation in space, the exact recognition of surroundings is essential [*Witkind and Wapner*, 1950]. Image slipping on the retina may result in disorientation in space. *Fukuda* [1957] reported that the fixation point between eyes and moving stripes of OK stimulator made dizziness much stronger. In this case the image slipping undoubtedly happened. Therefore increasing dizziness may be a sign of disorientation in space caused by this image slipping.

In this study, the body deviates to the same direction in all test conditions, except when the targets move slower than 45°/s in the usual OK stimulation.

Eye velocity in the look peripheral OK nystagmus is always slower than that of the target even in slow-speed stimulation [*Miyoshi* et al., 1978]. Eye velocity of the stare OK nystagmus is slower than that of look OK nystagmus [*Honrubia* et al., 1939]. In both cases of peripheral OK nystagmus, look and stare, the images of the stripes slip on the retina.

When the subject fixes his eyes upon the LED on the screen, the image slipping of the stripes undoubtedly occurs. The common phenomena in the above-mentioned several cases in which the body center of gravity deviates to the same direction of stimulation is the image slipping of surroundings on the retina. The subject cannot recognize the circumstances exactly, owing to this image slipping and looses his orientation in space. Thus the body is attracted by the movement of surroundings and deviates to the same direction of movement.

Conclusion

(1) The body sway induced by OK stimulation of whole visual field shows biphasic deviation. In the slow-speed range of stimulation, the body deviates to the counter-direction of rotation. In the high-speed range, on the contrary, the body deviates to the same direction of stimulation. (2) The former phase must be the coordinative one and the latter phase must be the disturbed phase. (3) Body sway induced by peripheral OK stimulation, regardless of look or stare, shows one-directional deviation, to the same direction of stimulation. (4) The body is attracted to the same direction of stimulation, too, when a fixed point is set on the screen of the OK stimulator. (5) The states of these latter three conditions are quite same as that of second phase of usual OK stimulation, that is to say, the disturbed one. (6) The disturbance of body balance of this kind arises from disorientation in space due to the image slipping on the retina.

References

Aschan, G.; Bergstedt, M.; Stahle, J.: Nystagmography. Acta Oto-lar., suppl. 127, pp. 1–103 (1955).
Fischer, M.H.; Veits, C.: Über optokinetisch ausgelöste Körperreflexe beim Menschen. Pflügers Arch. ges. Physiol. *219:* 579–587 (1927).

Fukuda, T.: Physiology of reflex in movement and equilibrium (in Japanese) (Igaku-shoin, Tokyo 1957).

Honrubia, V.; Downey, W.L.; Mitchell, D.P.; Ward, B.A.; Ward, P.H.: Experimental studies on optokinetic nystagmus. II. Normal humans. Acta oto-lar. Stockh. 65: 441–448 (1939).

Miyoshi, T.; Pfaltz, C.R.: Studies on the correlation between optokinetic stimulus and induced nystagmus. ORL 35: 350–362 (1973).

Miyoshi, T.; Shirato, M.; Hiwatashi, S.: Foveal and peripheral vision in optokinetic nystagmus; in Hood, Vestibular mechanisms in health and disease (Academic Press, London 1978).

Miyoshi, T.; Shirato, M.; Hiwatashi, S.: Two phasic body sway by optokinetic stimulation. Aggressologie 20: 119–125 (1979).

Vogel, P.: Über optokinetische Reaktionsbewegungen und Scheinbewegungen. Pflügers Arch. ges. Physiol. 228: 632–643 (1931).

Witkind, H.A.; Wapner, S.: Visual factors in the maintenance of upright posture. Am. J. Psychol. 63: 31–50 (1950).

T. Miyoshi, MD, Department of Otolaryngology, Faculty of Medicine,
Kyoto University, Kyoto 606 (Japan)

Adv. Oto-Rhino-Laryng., vol. 30, pp. 108–124 (Karger, Basel 1983)

Effect of Velocity, Amplitude (Stripe Distance) and Frequency of the Target on Eye-Tracking Test and Optokinetic Nystagmus

A. Tamada, M. Miyoshi, S. Ito, S. Hiwatashi

Faculty of Medicine, Kyoto University, Kyoto, Japan, and
Department of Otolaryngology, Osaka Red Cross Hospital, Osaka, Japan

Introduction

When a human subject follows a moving object with his eyes, the eyes perform two different types of movements: smooth pursuit movement and rapid, jerky changes of eye position, known as saccades. The eye tracking test (ETT) is designed to detect the accuracy of smooth pursuit movements; the optokinetic nystagmus (OKN) test measures the reflex mechanism of alternate eye movements of smooth pursuit and saccade.

ETT has usually been measured with instruments moving horizontally in a sinusoidal fashion such as a pendulum or a metronome. These devices have produced a smooth pursuit eye movement, without discontinuation, in a sinusoidal manner but the eye velocity thus produced changes with time; the distance between the subject and the target has never been standardized. Therefore ETT has been discussed only from the aspect of frequency but not from the aspect of velocity or amplitude; these other parameters have been difficult to quantify in ETT.

OKN consists alternately of a slow component and a rapid component. Recently, the slow phase has been considered to be identical to smooth pursuit eye movement and the rapid phase to saccade. OKN has, however, been reported to have two kinds of nystagmus. These are usually called 'foveal OKN' and 'retinal OKN'. In foveal OKN, the moving target is projected on the fundus in the retina and never errs from that small region, arcing about 1°. In contrast to this, in retinal

OKN, the moving object is not captured in the fundus but is projected mostly on the peripheral retina. The reason why these two kinds of OKN exist has never been clarified and the interrelationship between them is not well understood. As the slow phase in OKN usually has a constant velocity in ramp, it has been discussed mainly from the aspect of velocity but not from the aspects of amplitude (stripe distance) or frequency.

Thus the interrelationship among the frequency, amplitude and velocity of the target or stripe in ETT and OKN has never been elucidated. In order to clarify these interrelationships, we have used a device which produces a trigonal wave for the measuring of ETT. The characteristics of ETT thus obtained were compared to those in the slow phases of foveal and whole visual field OKN.

In addition, we have compared the interrelationship between the slow phase of foveal OKN and that of whole visual field OKN.

Methods

Subjects
10 normal individuals were studied in ETT and OKN respectively; no neurological deficits were recognized in their histories. Their ages ranged from 24 to 48 years. The sex ratio of female to male was 8 to 2 both in ETT and OKN.

Measurement of ETT
The subject, seated in a chair with the head mechanically fixed, and was instructed to follow a dot (visual arc: 1.2°) projected on a screen (height, 40 cm; width, 210 cm) which was placed about 1 m in front of the subject; the movement of the dot which moved at a constant velocity back and forth was recorded as a trigonal wave on DC-ENG polygraph.

Electro-oculograms of horizontal eye movements were differentially recorded with electrodes fixed lateral to both outer canthi and to the forehead as a reference ground for binocular recordings.

These binocular recordings from all 10 subjects were recorded on a DC-ENG polygraph; the frequencies (0.1, 0.3, 0.5, 0.8 and 1 Hz) with amplitudes of 10, 20, 30, 60 and 90° were made by changing the distance and velocity of the dot which moved horizontally on the screen.

In order to prevent habituation, all the subjects were examined only once at the first trial without any exercise beforehand.

Eye velocity was calculated on a DC-ENG record as shown in figure 1. Five continuous waves were randomly chosen and the velocity of two single waves (to the right and left) of each of the five waves were calculated. The velocities of these ten waves were added together and then divided by 10, resulting in the mean eye velocity for a subject at

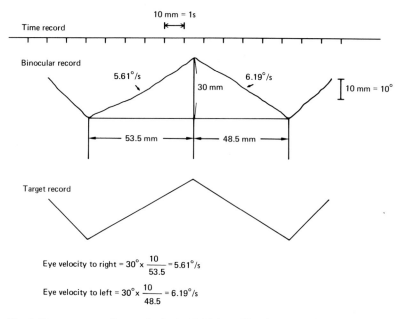

Fig. 1. Measurement of eye velocity in ETT. In calibration, a vertical deflection of 10 mm is equal to 10° of horizontal eye movement. In time record, 10 mm is equal to 1 s. Eye velocity is calculated as shown in the formula at the bottom of this figure.

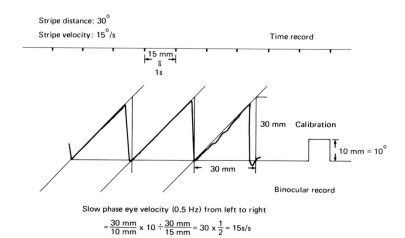

Fig. 2. Slow phase eye velocity in L-OKN. Measurement of eye velocity in OKN. A vertical deflection of 10 mm is adjusted to equal 10° of horizontal eye movement. In time record, 15 mm is equal to 1 s. Eye velocity of the slow phase is calculated as shown at the bottom of this figure.

each frequency with each amplitude. Then the mean eye velocities of all 10 subjects were added together and divided by 10, resulting in the final mean eye velocity at each frequency with each amplitude.

Measurement of OKN

To produce OKN, a projector-type OKN device was used. A light stripe was projected on a cylindrical white screen (radius, 90 cm; height, 200 cm) from a drum with slits in the center revolving around a light source. This revolving drum was a manually made metal cylinder (10 cm in diameter, 10 cm in height). The slits consisted of horizontal pin holes located every 10° in the center and was designed to project light stripes with widths of 1.7° on to the screen in a visual arc. These stripe distances were changed to 10, 20, 30, 60 and 90° by covering the pin holes with a black tape.

The drum was suspended at the center of the cylindrical screen just above the subjects head. The subjects were seated in a chair and were instructed not to turn their heads which were fixed to a chin platform while the stripe was rotating on the screen in the dark room.

The velocities of the light stripes were calculated for each frequency (0.3, 0.5, 0.8, 1, 2 and 3 Hz) at the stripe distances (amplitudes) of 10, 20, 30, 60 and 90°. First, the whole visual field OKN was recorded on DC-ENG polygraph at each velocity with each stripe distance. Second, only one stripe, designed to remain on the screen by covering the rest of the pin holes with black paper, was recorded on DC-ENG polygraph in a similar manner. The velocities of the slow phases of these OKN were supposed to be compatible with the velocities of the moving stripes. The velocities of these slow phases were calculated (fig. 2) and compared to each other and to those of ETT. Ten continuous slow waves were calculated for each frequency at each stripe distance; by dividing these values by 10, a mean eye velocity for each subject was obtained, then, these mean eye velocities from the 10 subjects were added together and divided by 10, resulting in a final mean eye velocity.

For the purpose of this experiment the one stripe OKN is considered to be the foveal OKN.

Results

Mean Eye Velocity in ETT

Mean eye velocities of 10 normal subjects calculated at each frequency for each amplitude in ETT are presented in table I.

The relationships between eye velocity and target velocity for each amplitude are depicted in figure 3. The discrepancy between the target velocity and the eye- following velocity becomes larger as the target velocity increases. At the small amplitudes, this occurs even at a low target velocity while at large amplitudes, the same degree of discrepancy occurs at a high target velocity. However, the maximum eye velocity at which the human eye can smoothly follow a target appears to be

Table I. Mean eye velocity in ETT

Amplitude	Frequency Hz	Eye velocity °/s	Target velocity °/s	Eye velocity/ Target velocity
10°	0.1	1.981	2	0.991
	0.3	5.684	6	0.947
	0.5	8.876	10	0.8876
	0.8	12.681	16	0.7926
	1.0	15.489	20	0.7745
20°	0.1	3.978	4	0.9945
	0.3	11.204	12	0.9336
	0.5	17.820	20	0.891
	0.8	25.651	32	0.8016
	1.0	30.382	40	0.7596
30°	0.1	6.019	6	1.003
	0.3	17.737	18	0.9855
	0.5	26.597	30	0.8866
	0.8	38.272	48	0.7973
	1.0	45.279	60	0.7547
60°	0.1	11.965	12	0.9970
	0.3	33.443	36	0.9289
	0.5	52.480	60	0.8747
	0.8	71.881	96	0.7488
	1.0	86.212	120	0.7184
90°	0.1	17.950	18	0.9972
	0.3	50.279	54	0.9311
	0.5	75.987	90	0.8442
	0.8	109.754	144	0.7621
	1.0	128.600	180	0.7144

around 60°/s; at greater velocities, the eye can no longer follow the target smoothly even at wide amplitudes.

Within this maximum eye velocity, however, we can say that as the amplitude becomes wider, the human eye is able to follow a target at a faster velocity without help from saccade. But as the range at which the human eye can conjugately move is considered to be within a visual arc of about 120°, the influence of this limitation appears only when the amplitude is more than 90°.

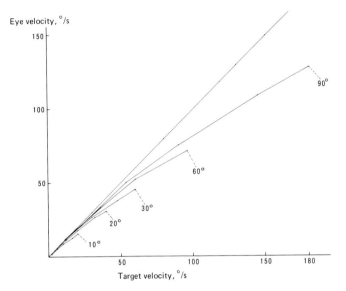

Fig. 3. Eye velocity is plotted on the ordinate and target velocity on the abscissa in ETT. At small amplitudes, the eye becomes unable to follow a target even at a low velocity, though it can follow a target at a faster velocity at a large amplitude.

In figure 4, the mean eye velocity/target velocity ratio is calculated and plotted on the ordinate and target velocity on the abscissa in order to assess the degree of difficulty in following the target at each amplitude. As figure 4 shows, the ratio at narrow amplitudes (10, 20 and 30°) becomes remarkably small, even when the target velocity is low. In contrast, this ratio at large amplitudes (60 and 90°) is still near 1 even when the target goes faster, up to about 60°/s. Figure 4 clearly shows that ETT is influenced not only by the velocity but also by the amplitude of the target.

The relationship between frequency and amplitude of the target in ETT is depicted in figure 5. The ratio of mean eye velocity to target velocity at each amplitude progressively decreases as the frequency of the target becomes larger, also the degree of the progressive decrement of that ratio is nearly identical at all amplitudes. This means that the human eye can follow a target up to 0.5–1 Hz smoothly regardless of the amplitudes applied and this ability is influenced remarkably by the frequency.

These findings indicate that ETT is mainly influenced by the frequency but not by the velocity or amplitude of the target.

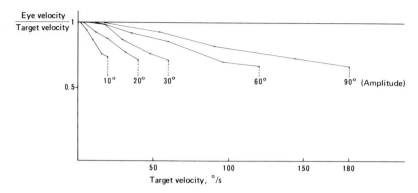

Fig. 4. The ratio of eye velocity to target velocity is plotted on the ordinate, and the target velocity on abscissa in ETT. At small amplitudes, the ratio becomes remarkably smaller at slow target velocities; while at large amplitudes, this is not so.

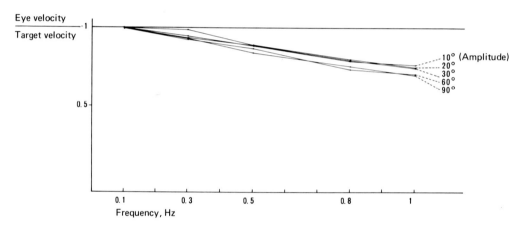

Fig. 5. The ratio of eye velocity to target velocity is plotted on the ordinate, and the frequency on abscissa in ETT. Regardless of the amplitudes, the ratio at each amplitude decreases gradually, as the frequency increases.

Mean Eye Velocity in Slow Phase of OKN

Eye velocities of the slow phases in the whole visual field OKN and foveal OKN (strictly speaking, one stripe) are shown in table II for each frequency at each stripe distance (amplitude). The ratios of eye velocity to stripe velocity are also shown for the whole visual field OKN and foveal OKN.

Table II. Slow phase velocity in L-OKN

Stripe distance	Frequency Hz	Stripe velocity °/s	Eye velocity, °/s[1]		Eye velocity/ Stripe velocity	
			whole visual field	foveal	whole visual field	foveal
10°	1.0	10	10.3 ± 2.18	9.6 ± 1.58	1.03	0.96
	1.5	15	14.6 ± 1.39	13.7 ± 1.80	0.98	0.93
	2.0	20	19.3 ± 1.41	16.3 ± 1.82	0.97	0.82
	3.0	30	29.1 ± 1.32	23.8 ± 3.28	0.97	0.79
20°	0.5	10	10.4 ± 1.41	8.8 ± 2.00	1.01	0.83
	0.8	16	16.6 ± 1.94	14.7 ± 1.78	1.03	0.90
	1.0	20	16.6 ± 1.94	14.7 ± 2.03	1.02	0.89
	1.5	30	28.8 ± 2.90	25.1 ± 2.56	0.95	0.83
	2.0	40	38.9 ± 3.64	31.2 ± 4.22	0.97	0.76
	3.0	60	53.0 ± 3.16	34.5 ± 10.9	0.89	0.54
30°	0.3	9	8.9 ± 1.90	8.7 ± 1.10	0.97	0.94
	0.5	15	15.3 ± 1.92	14.9 ± 1.49	0.98	0.88
	0.8	24	24.2 ± 2.92	23.6 ± 3.03	0.99	0.98
	1.0	30	29.1 ± 3.06	28.8 ± 2.74	0.95	0.95
	1.5	45	42.0 ± 2.67	39.3 ± 4.49	1.05	0.86
	2.0	60	54.9 ± 6.04	46.4 ± 6.34	0.91	0.76
	3.0	90	67.4 ± 8.46	43.1 ± 17.82	0.74	0.45
60°	0.3	18	18.6 ± 3.01	17.0 ± 1.12	1.04	0.94
	0.5	30	31.9 ± 3.71	28.2 ± 2.59	1.04	0.94
	0.8	48	46.8 ± 3.34	44.1 ± 3.32	0.96	0.92
	1.0	60	55.5 ± 7.84	51.7 ± 7.19	0.91	0.84
	1.5	90	79.8 ± 7.01	67.9 ± 13.06	0.88	0.73
	2.0	120	81.1 ± 13.16	66.7 ± 18.95	0.69	0.53
	2.5	150	81.4 ± 7.27	57.6	0.54	0.38
90°	0.1	9	8.9 ± 0.72	7.7 ± 1.23	0.99	0.86
	0.3	27	27.2 ± 2.97	25.7 ± 2.41	1.01	0.95
	0.5	45	44.6 ± 5.35	42.8 ± 3.26	0.99	0.95
	0.8	72	66.2 ± 7.79	62.8 ± 6.06	0.92	0.87
	1.0	90	77.9 ± 10.42	70.7 ± 8.56	0.87	0.79
	1.5	135	93.1 ± 12.24	81.7 ± 10.44	0.69	0.61
	2.0	180	74.6 ± 26.18	54.2 ± 11.4	0.41	0.30

[1] Mean ± SD.

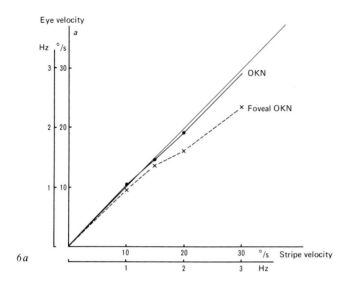

6a

Mean Eye Velocity in Slow Phase of Foveal OKN (One Stripe)

The relationship between eye velocity and stripe velocity is depicted at the stripe distances of 10, 20, 30, 60 and 90° in figure 6 (dotted
line). As the stripe distance becomes wider, the eye can smoothly follow a moving stripe at faster velocities up to about 60°/s. Stated in
terms of frequency, the eye can follow a stripe up to about 1 Hz regardless of the stripe distance applied. At frequencies greater than 1 Hz, the
difference between eye velocity and stripe velocity becomes larger, and
at a stripe distance of 90°, this difference can be seen at 0.8–1 Hz because the limitation in eye movement and the maximum eye velocity of
humans become apparent at this amplitude. The ratio of eye velocity to
stripe velocity (fig. 7) indicates that as the stripe distance becomes
wider, the eye can follow the stripe at a faster velocity, thus, the stripe
distance is an influence on the velocity of the slow phase in foveal
OKN. Figure 8 shows the influence of frequency on foveal OKN. The
human eye can follow any stripe up to 1 Hz. The slow phase of foveal
OKN is influenced not only by the velocity but also by the stripe distance (amplitude) of the target (fig. 6–8). Moreover, the parameter of
frequency seemed to be the most important factor in measuring foveal
OKN, as figure 8 shows.

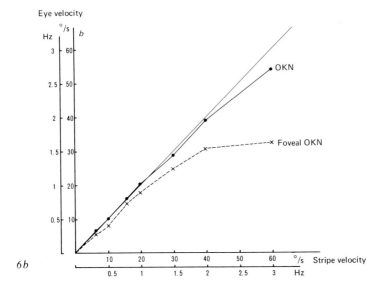

6b

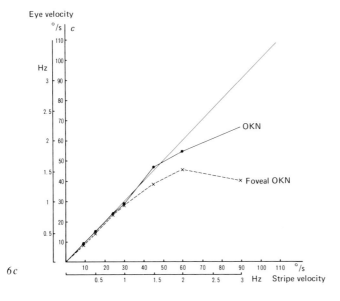

6c

Fig. 6. The relationships between eye velocity and stripe velocity (with frequency) are depited in a slow phase of foveal OKN (----) and that of the whole visual field OKN (——). As the amplitude increases, the maximum eye velocity also increases. It is about 60°/s in foveal OKN, while it is around 70–80°/s in whole visual field OKN. In foveal OKN, the eye can follow a target smoothly up to 1 Hz, while in whole visual field OKN it can follow up to 1.5–2 Hz. L-OKN in stripe distances: (a) 10°; (b) 20°; (c) 30°; (d) 60°, and (e) 90°.

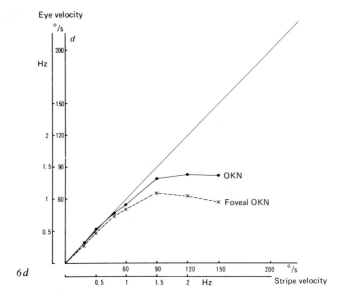

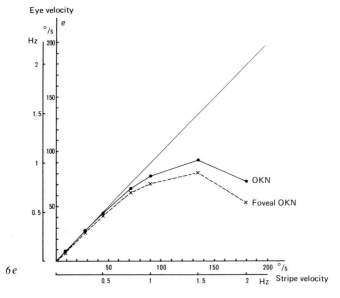

For legend to figure 6 see reverse side.

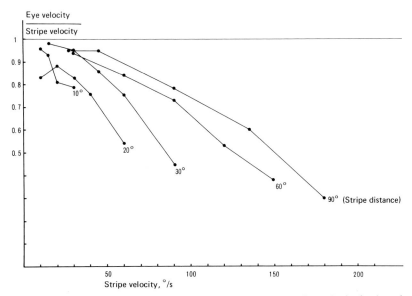

Fig. 7. L-OKN in foveal vision. The ratio of eye velocity to stripe velocity is plotted on the ordinate and stripe velocity on the abscissa in foveal OKN. As stripe distance (amplitude) increases, the maximum eye velocity for following a target also increases.

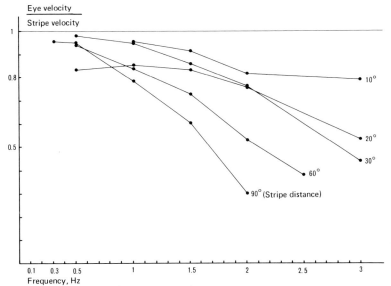

Fig. 8. L-OKN in foveal vision. The ratio of eye velocity to stripe velocity is plotted on the ordinate and frequency on the abscissa in foveal OKN. Up to 1 Hz, the eye can follow a target smoothly at all the amplitudes.

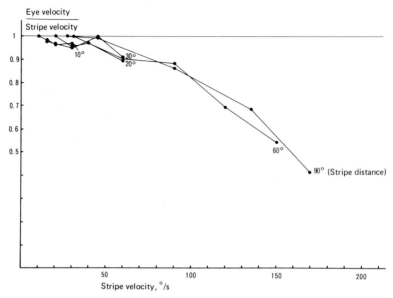

Fig. 9. L-OKN in whole visual field. The ratio of eye velocity to stripe velocity is plotted on the ordinate and stripe velocity on the abscissa in whole visual field OKN. The decrease in the ratio is similar at all amplitudes, when the velocity increases.

Mean Eye Velocity in Slow Phase of Whole Visual Field OKN

The relationship between eye velocity and stripe velocity at stripe distances of 10, 20, 30, 60 and 90° is depicted in figure 6 (solid line). The values obtained for the slow phase of the whole visual field OKN were always higher than those of foveal OKN, indicating that the whole visual field OKN permits the human eye to follow a stripe at a faster velocity smoother than the foveal OKN when stripe velocity becomes faster. The maximum eye velocity to follow the stripe smoothly in this condition seems to be about 70–80°/s. The eye can follow a stripe up to 1.5–2 Hz, however, at a stripe distance of 90° the frequency is limited to around 1 Hz. The degree of difficulty for the human eye to follow the stripe in this condition is shown in figure 9 where the eye velocity/stripe velocity ratio is plotted against stripe velocity. A progressive decrease in these ratios at all stripe distances is evident.

Figure 9 also shows that as the stripe velocity increases, the ability of the human eye to follow the stripe decreases but this degree of discrepancy is not as great as that seen in foveal OKN. Figure 10 shows

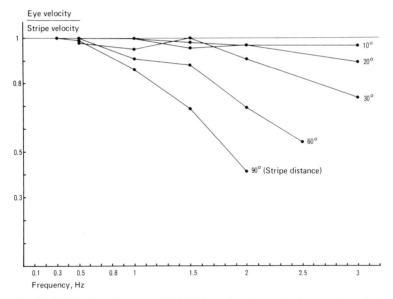

Fig. 10. L-OKN in whole visual field. The ratio of eye velocity to stripe velocity is plotted on the ordinate and frequency on the abscissa in whole visual field OKN. The eye can follow a target up to 1.5–2 Hz except at a stripe distance of 90°.

that the eye can follow a stripe smoothly up to 1.5–2 Hz at stripe distances of 10, 20° and 30° but not at a stripe distance of 60 or 90°. Even with respect to frequency, the ability of the human eye to follow the stripe is greater than that in foveal OKN.

Interrelationship between
Slow Phases in Foveal OKN and in Whole Visual Field OKN

In the slow phase of foveal OKN, the tendency for the human eye to follow a target is almost the same as that seen in ETT. In contrast to this phenomenon, the human eye can follow a target at a faster velocity and higher frequency in the whole visual field OKN (fig. 9, 10) than in either foveal OKN (fig. 7, 8) or ETT.

This probably means that as the numbers of stripes projected on the peripheral field of the retina increase, the human eye can see the target more clearly and follow it more smoothly at a faster and higher frequency.

Normally in the whole visual field OKN, peripheral vision is added to the foveal vision, assisting the latter greatly.

Discussion

In human subjects, two kinds of OKN have been reported by *Ter Braak* [1], *De Kleyn* [2], *Jung* [3] and *Honrubia* et al. [4] and the numerous nomenclatures for these OKN were thereafter proposed. *Jung* [3] described these OKN as foveal and retinal nystagmus. In the former, objects are always being projected on the fundus in the retina and in the latter they are seen primarily on the peripheral retinal areas.

Miyoshi et al. [5] distinguished foveal nystagmus from peripheral nystagmus and discussed in detail the interrelationships between them, stating that both are necessary to bring about smooth OKN in human beings.

In this paper, we investigated the interrelationships between the slow phase in foveal OKN and that in whole visual field OKN in order to clarify the role played by the peripheral vision in OKN. In addition, the slow phases of these OKN were compared to ETT. The mutual interrelationships among velocity, amplitude (stripe distance) and frequency of the target which influence ETT and OKN were also discussed.

As figure 6 shows, the slow phase in the whole visual field OKN (solid line) is different from that in foveal OKN (dotted line). The neural pathways in these OKN have not been fully understood in regard to whether they have a common neural pathway or different pathways. As far as our experiments were concerned, the slow phase in the whole visual field OKN was more reflexive than in the foveal OKN. The characteristics of the latter seemed to be similar to those seen in ETT because, in both, the fundus in the retina plays a major role in the following of a target, suggesting that a common neural pathway might be involved.

In contrast, the slow phase in the whole visual field OKN was considered to be more sophisticated than in the foveal OKN or ETT. The so-called 'smooth pursuit eye movement' usually refers to those seen in the latter group. We will discuss the relationships among the parameters on OKN and ETT, focusing primarily on the latter group.

In ETT and OKN, the relationships among the velocity, frequency and amplitude (stripe distance) of a target have never been clarified and therefore it has been difficult to make quantitative measurements of ETT and OKN, although there have been many qualitative interpretations. For this reason there has been much confusion in making a correct diagnosis using ETT for the patients with neurological deficits. In order to avoid these defects, we used a constant trigonal wave in ETT instead of the sinusoidal wave in which velocity is never constant.

As figures 3–5 show in ETT, the parameter of frequency was the most influential in measuring ETT, compared to the other parameters such as velocity or amplitude. The relationships of stripe velocity or frequency to the ability of the human eye to smoothly follow a target are depicted in figures 7 and 8 in foveal OKN. These figures are similar to those seen in figures 4 and 5 in ETT. Figure 3 also demonstrates a tendency similar to that seen in figure 6 in foveal OKN with regard to velocity, indicating that the slow phase in foveal OKN is similar to ETT and is mainly influenced by the frequency but not by the velocity or distance itself of the stripe.

In regard to the slow phase of the whole visual field, the influence of frequency is greatest up to 70–80°/s of the target velocity but higher than that, velocity becomes the main influencing factor probably because the limitations of the maximum eye velocity and eye movement have to be taken into account in this condition (fig. 9, 10).

Although *Baloh* et al. [6] and *Yamazaki* [7] discussed the quantitative measurement of ETT using a constant trigonal wave, they were only concerned with velocity, and not with the effects of amplitude or frequency. They were able to demonstrate that as the velocity of a target increases, smooth pursuit eye movement becomes jeopardized. However, they did not report amplitude related to changes in the maximum eye velocity. ETT is influenced constantly by the frequency, as shown in figures 3–5, regardless of the changes in velocity or amplitude of the target. From these facts it might be better to discuss ETT in terms of frequency and not velocity or amplitude itself. The same can be said of the slow phase of foveal OKN which is influenced not only by velocity but also by stripe distance, namely frequency. Under similar conditions, the frequency and velocity are prominently improved in the slow phase of whole visual field OKN as shown in figures 9 and 10, compared to those in foveal OKN. We postulate that this phenomenon is due to the contribution that peripheral vision makes to the foveal vision. Namely, peripheral vision assists foveal vision and therefore the human eye is able to follow an object moving at a faster velocity and higher frequency. In this slow phase, the reflex in peripheral vision on foveal vision is at work; this is different from the ordinary smooth pursuit eye movement shown in ETT or the slow phase of foveal OKN.

This indicates that if the stripes are numerous in the peripheral retina, the resolution of the human eye increases and thus can follow an object more accurately. As the region stimulated in peripheral retina becomes wider, the human eye can follow a target at a faster and higher frequency. We would like to call this phenomenon 'peripherally assisted foveal vision'.

References

1   Ter Braak, G.: Untersuchung über optokinetischen Nystagmus. Arch. Neerl. Physiol. *21:* 309–376 (1936).
2   De Kleyn, A.: The connection between the optokinetic nystagmus and the vestibular system. Acta oto-lar., suppl. 78, pp. 8–13 (1949).
3   Jung, R.: Handbuch der inneren Medizin; 4. Aufl., vol. 5/1: Neurologie, p. 1325 (Springer, Berlin 1953).
4   Honrubia, V. et al.: Experimental studies on optokinetic nystagmus. II. Normal humans. Acta oto-lar. *65:* 441–448 (1968).
5   Miyoshi, T. et al.: The meaning of the position and wideness of the visual defect in eye movement; in Suzuki, Brain and disequilibrium, pp. 91–118 (Shinoharashupan, Tokyo 1981).
6   Baloh, R.W. et al.: Quantitative measurement of smooth pursuit eye movements. Ann. Otol. Rhinol. Lar. *85:* 111–119 (1976).
7   Yamazaki, A.: Quantitative measurement of smooth pursuit eye movement. Clin. Ophthal. Jap. *31:* 47–50 (1977).

A. Tamada, MD, Kyoto University, Faculty of Medicine, Kyoto 606 (Japan)

Adv. Oto-Rhino-Laryng., vol. 30, pp. 125–127 (Karger, Basel 1983)

Optokinetic Nystagmus Pattern in Subjects Exposed to Prolonged Local Vibrations

A.R. Halama

Department of Otolaryngology, Academy of Medicine, Krakow, Poland

Several authors [1, 3, 6, 8] have already stressed the usefulness of caloric and rotational tests in the diagnosis of vibrations disease, although the results were not very consistent [8]. On the contrary, not much attention has been focused on the optokinesis and its disturbances as a possible cause of the above-mentioned complaints in this disease.

The aim of this report is to present our optokinetic findings obtained in a homogeneous professional group of locksmith-riveters chronically exposed to local vibrations generated by so-called 'stroke handtools', i.e., mechanical repeated stroke riveting tools and electric handdrills.

Material and Methods

The optokinetic test as a part of the hearing and vestibular function evaluation was performed in 338 locksmith-riveters employed in an airplane factory. The age of subjects varied from 25 to 60 years. During an 8-hour shift each subject drills aluminium sheets for approximately 4 h, and performs riveting for 4 h. Both operations are carried out by using electric handdrills and special stroke riveting tools which generate vibrations exceeding 4 times the admissible vibration standards in octaves with central frequencies of 63 and 125 Hz. The subjects have been working in this same environment from 5 to 25 years.

The stimulation was alternatively applied, first to the left, then to the right with at least a 3-min break between consecutive stimulations. The stimulus velocity was 30, 60, 90°/s. Although the ENG recording time was 20 s, the 10 s periods were evaluated. For the quantitative assessment of the optokinetic reaction the difference between left and right responses was calculated as the percentage of the total response. For this purpose the Jongkees formula was adopted [2]. We accepted a difference between left and right as pathological as regards the average velocity of the slow phase when it exceeded the value by 25%. In addition, the qualitative assessment of the optokinetic response was also performed: regularity, persistence, presence of particular nystagmus phases.

Results

Among 338 subjects examined, only 92 cases had symmetrical responses to all stimulations obtained and no disturbed nystagmus patterns were seen. In the remaining 246 cases, the optokinetic traces showed a variety of quantitative and qualitative abnormalities. In this group, 67 cases of asymmetry of responses but with undisturbed nystagmus pattern were found to all consecutive stimulations. 75 cases, however, who showed symmetrical responses also presented a dissociation of the nystagmus pattern with short periods of nystagmus decline up to complete disappearance or occurrence of beats with indistinguishable phases. This phenomenon was more pronounced with the stimulus duration as well as with increase in velocity. In the other 68 cases, the response for 30°/s was symmetrical: nevertheless, for the consecutive stimuli 60 and 90°/s the responses were asymmetrical even though the examination was repeated several times in different sessions.

In 36 cases we were unable to elicit optokinetic reactions to all three pairs of stimulations, although we had been testing these subjects on repeated occasions. In these cases direct observation did not reveal any clear nystagmus either. It is worth mentioning that the majority of subjects with abnormalities in optokinetic recordings during the stimulations complained about dizziness, spatial disorientation, often with clear vegetative reactions. We did not observe this during vestibular tests.

Discussion

Several optokinetic abnormalities of diagnostic importance have been described [4, 5, 7, 9, 10]. When evaluating our optokinetic findings it can be stated that a variety of already described abnormalities were present in 73 % of subjects exposed to local as opposed to general vibrations. But considering all known obstacles in obtaining reliable obtokinetic recordings, certain doubt can be expressed as to whether part of them are not simple artefacts. But this objection can be rejected on the grounds that our testing procedure was standardized and that the results were reproducible. Therefore, the following conclusion can be drawn from our results, i.e. that in addition to the already described abnormalities in the CNS and vestibular system due to intense and prolonged local vibration exposure [1, 6, 8] severe disturbances of optokinesis can also occur. Such frequent occurrence of optokinetic abnormalities in this particular professional group seems to be well substantiated if we consider the specifity of the operations, prolonged eye fixation on the vibrating target, duration of local vibration, type of vibration. At the present stage of our investigation, we cannot answer whether these optokinetic abnormalities are functional or due to organic lesions of the optomotoric subsystem. But whatever the specific pathomechanism of these abnormalities is, it is probable that certain complaints reported by subjects exposed to prolonged local vibrations which are not objectivated in view of vestibular findings, may be related to disturbed optokinesis.

References

1   Andrejewa-Galanina, E.C.: Higieniczne znaczenie odrzutów i wibracji młotków pneumatycznych. Med. Pracy 7: 5–339 (1957).
2   Bień, S.; Żmigrodzka, K.: W sprawie ilosciowej oceny oczoplasu opto kinetycznego. Otolar. pol. 30: 579–586 (1976).
3   Bochenek, Z.: Otoneurologia kliniczna, pp. 3–304 (PZWL, Warszawa (1977).
4   Carmichael, E.; Dix, M.; Hallpike, C.S.: Lesions of the cerebral hemispheres and their effect upon optokinetic and caloric nystagmus. Brain 77: 345–372 (1954).
5   Cogan, D.G.: Neurology of the ocular muscles, pp 3–296 (Thomas, Springfield 1966).
6   Grzegorczyk, L.; Waleszek, M.: Drgania i ich oddzialywanie na organizm ludzki, pp. 5–320 (PZWL, Warszawa 1972).
7   Jung, R.; Kornhuber, H.H.: Results of electronystagmography in man: the value of optokinetic vestibular and spontaneous nystagmus for neurological diagnosis and research, pp 4–311 (Harper & Row, New York 1964).
8   Kieszkowska, L.: Badania narzadu sluchu i rownowagi u osob narazo nych na dzialanie wibracji. Otolar. pol. 22: 233–235 (1978).
9   Maspetiol, R.; Semette, D.; Jachowska-Hederman, A.: L'étude du nystagmus optocinèque en oto-neurologie Annls. Oto-lar. 85: 5–24 (1968).
10 Niederding, P.H.: Optokinetic nystagmus: a comparative study of two stimulators. Laryngoscope 89: 779–793 (1979).

A.R. Halama, MD, PhD, KNO Service – VUB, Laarbeeklaan 101,
B-1090 Brussels (Belgium)

Adv. Oto-Rhino-Laryng., vol. 30, pp. 128–130 (Karger, Basel 1983)

Optokinetometry:
A Diagnostic Tool in Multiple Sclerosis

J. Láng[a], *A.F. Mester*[b]

[a] Department of Neuro-Otology, Korányi Hospital, Budapest, Hungary, and
[b] Department of Otolaryngology, Semmelweis Medical School, Budapest, Hungary

Multiple sclerosis (MS) is a multifocal, periodical, progressive, chronic disease of the white matter of the central nervous system. In the case of a single symptom, the detection of a latent focal sign may give a decisive hint towards the diagnosis. In a previous paper [*Láng and Orbán*, 1964] we reported on the electronystagmographic (ENG) findings of caloric nystagmus in patients with a just beginning, demyelinating process. In this paper we are concerned with clinically developed MS patients.

Material and Methods

On 100 in-patients (30 men and 70 women), at the Department of Neurology and Psychiatry of the Medical School Pécs, being treated for MS, we performed the complete conventional battery of tests for vestibular and optokinetic function. Optokinetic nystagmus (OKN) was elicited with a foveal optokinetic stimulator developed by *Láng* [1962]. Recordings were made with simultaneous monocular and binocular registration with horizontal and vertical leads on a 16-channel Medicor ELFI Electronystagmograph. Technical details were presented in a previous paper [*Mester and Pálffy*, 1981].

Results

Of the 100 MS patients, 17 were cases with ophthalmologically established, clinically manifest binocular bilateral nasal slowing (INO). In the optokinetogram, a dissociation of the response, ataxic over-

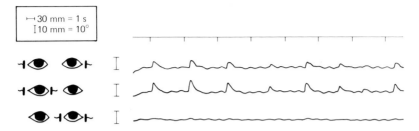

Fig. 1. Clinically manifest internuclear ophthalmoplegia. Optokinetic nystagmus: 20°/s. Binocular, right and left monocular channels.

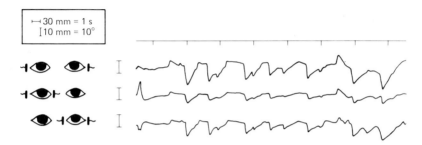

Fig. 2. Subclinical internuclear ophthalmoplegia. Optokinetic nystagmus: 10°/s. Binocular, right and left monocular channels.

shoots and stepped, broken-up pursuit movements, especially manifest in the eye-tracking test, are to be seen. Figure 1 shows the findings in a typical case of this group. In the binocular and right monocular trace (1st and 2nd channels) there is a discrepancy between target speed and eye velocity. The left eye trace shows no nystagmic response, but a fine oscillation of waves. In 13 cases in which INO was subclinical, the optokinetogram has discovered its latent presence (fig. 2). In the binocular trace the following movements are deranged, stepped and broken up. Comparing the abducting eye (3rd channel) with the adducting eye (2nd channel), the amplitude of the latter is lower, and the fast phase is less steep, and the angle between fast and slow phase is rounded. There is a definite lag of the adduction saccadic movement.

In several cases under steroid treatment, the ENG recording was performed repeatedly, and on the optokinetogram we could follow the slow regression of INO, which in one case was the last symptom to recover during remission.

Conclusions

We may conclude that deranged OKN is a frequent finding in MS [*Dix*, 1980; *Láng and Csapody*, 1973] which, by analyzing the foveally elicited and simultaneously recorded (monocular and binocular) tracings, reveals the presence of INO. It enables even the detection of its subclinical latent form. Thus optokinetometry may be useful in the diagnosis and follow-up of MS.

Acknowledgements

We gratefully acknowledge Prof. *G. Pálffy* who made available the patients upon which this study is based. Thanks are due to Mrs. *F. Tringer* for her skilful technical assistance.

References

Dix, M.R.: The mechanism and clinical significance of optokinetic nystagmus. J. Laryngol. *94:* 845–864 (1980).

Láng, J.: L'évocation du nystagmus opto-cinétique à l'aide d'un nouvel appareil et l'importance diagnostique de la méthode. Acta oto-rhino-lar. belg. *16:* 5–13 (1962).

Láng, J.; Csapody, I.: Standardized optokinetometry as part of the routine vestibular examination. Proc. 10th. Wld Congr. ORL, pp. 107–112 (1973).

Láng, J.; Orbán, L.: Zur Frühdiagnose der multiplen Sklerose aus otoneurologischer Sicht. Mschr. Ohrenheilk. Lar.-Rhinol. *98:* 502–505 (1964).

Mester, A.F.; Pálffy, G.: Detection of silent locus of patients with multiple sclerosis by the help of electronystagmography (in Hungarian) Orv. Hétil. *122:* 9–11 (1981).

J. Láng, MD, Department of Neuro-Otology, Korányi Hospital,
H-1074 Budapest (Hungary)

Adv. Oto-Rhino-Laryng., vol. 30, pp. 131–137 (Karger, Basel 1983)

Experiences with Neuro-Otological Procedures in the Diagnosis of Intracranial Pathology

J. Helms, M.Y. Abdel Aziz, K. Maurer

Universitäts-HNO-Klinik (Prof. Dr. *J. Helms*) and Psychiatrische Klinik
(Dir.: Prof. Dr. *O. Benkert*), Mainz, FRG

The aim of this study was to compare the diagnostic procedures frequently [1] used for the detection of intracranial pathology, with special respect to their value as screening tests. The main interest was acoustic neuromas as the major type of tumor in the posterior cranial fossa.

Koppenburg [5] analyzed the data of 1,029 patients suffering from sensorineural hearing loss of different etiology. This assay, obtained from our material in Prof. *Plester's* clinic in Tübingen, and data from Mainz were the basis for this presentation. Sudden deafness and Menière's disease were the most frequent diagnoses differentiating from acoustic neuromas. 81 out of 128 operated cases can be included in this series.

Pure subjective symptoms as tinnitus or the feeling of fullness were not included in the collection of these data.

The alternate binaural loudness balance test from Fowler shows approximately equal parts of positive recruitment signs in the three mentioned diseases with a slight dominance in Menière's disease (table I).

The short increment sensitivity index test was performed at 3 frequencies. The test was defined as positive in cases with a score over 75% and negative below 35%. The scores in between were considered uncertain. The results obtained depended on the test frequency. At 4,096 Hz acoustic neuromas showed positive and negative results at equal parts. When the results were compared to the nonaffected side, no difference was seen. This was the same in Menière's disease and sudden deafness (table II). The tone decay test showed normal adap-

Table I. Alternate binaural loudness balance test

	n	Positive, %	Negative, %	Not possible, %
Sudden deafness	140	64	5	31
Ménière's disease	119	76	1	23
Acoustic neuroma	35	65	3	32

Table II. Short increment sensitivity index (SISI) (measured at 4,096 Hz)

	n	Positive, %	Negative, %	Uncertain, %
Sudden deafness	68	63	25	12
Ménière's disease	63	68	21	11
Acoustic neuroma	18	45	45	11
Not affected ear	36	42	45	14

Table III. Tone decay test

	n	Adaptation normal, %	abnormal, %	unreliable, %
Sudden deafness	92	80	5	15
Ménière's disease	70	90	7	3
Acoustic neuroma	25	60	16	24

tion in 80% of patients with sudden deafness, 90% of patients with Ménière's disease and remained by 60% in acoustic neuromas (table III). Speech discrimination audiometry was more or less reflecting the pure tone hearing loss and was similar in all three mentioned diseases.

The vestibular tests by Romberg and Unterberger, and the gait test were normal in 60–90% of the cases. In the gaze tests 95% of the patients with sudden deafness, 94% of the Ménière's patients and 80% of all tumor patients showed normal results. The caloric test was evaluated for the warm phase, because this test gave the best results. This test was false-negative in 25% of surgically proven acoustic neuromas (table IV–VI). A positiogram was not performed.

Table IV. Vestibular function tests

	n	Falling to lesion side	No falling	Falling to normal side
Romberg				
Sudden deafness	209	2	205	2
Ménière's disease	142	3	137	2
Acoustic neuroma	43	2	40	1
Gait test				
Sudden deafness	204	19	174	11
Ménière's disease	138	13	115	10
Acoustic neuroma	42	9	29	4
Unterberger Test				
Sudden deafness	208	33	133	42
Ménière's disease	139	27	88	24
Acoustic neuroma	43	19	15	9

Neurological tests for the detection of trigeminal deficits in the face, or the Hitselberger test were found to be positive in 39% of the patients with acoustic neuromas. The protein concentration in the cerebrospinal fluid was increased to over 45 mg% in tumor cases in 50%.

The radiological examination of the temporal bone showed pathological changes in acoustic neuromas and also in cases with sudden deafness and Ménière's disease. The plain tomography did not prove to have a higher score in detecting an acoustic neuroma than the Stenvers projection (table VII). The so-called 'empty' internal auditory meatus will not be discussed here [2 4].

The cisternography revealing the tumor in 95% of the cases was the most important test in this series. In 5% it showed either a false-negative result or has been misinterpreted. An improvement was achieved by computerized tomography (CT).

59% of the acoustic neuromas did not show in plain CT and 27% were not detected after intravenous application of a contrast medium. CT combined with gas cisternography as described by *Sortland*[7] and by *Wende*[8], our neuroradiologist, ist the best technique at present for revealing intrameatal tumors. In the hands of the experienced neuroradiologist it is a conclusive and less confusing study than all other diagnostic tests (table VIII).

Table V. Vestibular function gaze tests

	Nystagmus					
	to lesion side	none	to normal side	to lesion side	none	to normal side
	Looking upwards			Looking downwards		
Sudden deafness	6	202	5	4	203	7
Ménière's disease	8	134	2	4	138	2
Acoustic neuroma	3	41	0	5	40	1
	Looking to lesion side			Looking to normal side		
Sudden deafness	5	202	7	4	206	3
Ménière's disease	8	135	1	6	136	2
Acoustic neuroma	7	37	1	4	38	3
	Looking forward					
Sudden deafness	4	205	8			
Ménière's disease	8	134	3			
Acoustic neuroma	4	41	1			

Table VI. Vestibular function: Caloric test with warm water

	n	Prolonged response, %	Normal activity, %	Hypoactivity, %
Sudden deafness	207	20	55	25
Ménière's disease	142	15	34	51
Acoustic neuroma	80	0	22	78

Since 1978 we have been using brain stem audiometry in the detection of acoustic neuromas. Since the tumor originated between the origin of wave I and II, we aimed at finding a stimulation technique which permitted the specific recording of these two waves. The method applied has been published by *Maurer* et al. [6]. With this method, it was possible to exclude those patients suffering from multiple scleroses, when the latency between wave I and II was normal and wave V was pathological.

Table VII. Radiology

	n	Normal	Pathological in %		
			lesion side	normal side	both sides
Stenvers projection					
Sudden deafness	210	94	3	1	2
Ménière's disease	144	92	6	1	2
Acoustic neuroma	81	24	72	0	4
Plain tomography					
Sudden deafness	25	92	8	0	0
Ménière's disease	32	81	19	0	0
Acoustic neuroma	72	16	78	3	3
Cisternography					
Sudden deafness	18	89	11	0	0
Ménière's disease	24	88	12	0	0
Acoustic neuroma	64	5	95	0	0

Table VIII. Computerized tomography: Results in acoustic neuromas

	n	Normal, %	Pathological, %	Uncertain, %
CT	29	59	31	10
CT with CM	30	27	60	13
CT with gas	10	0	100	0

2 patients showed a double peak of wave II, the first peak with normal, the second peak, with pathologic latency. This change could have been caused by slight pressure on the VIIIth nerve. Function preserving surgery revealed a nontumor pressure to the VIIIth nerve.

In 79 patients with acoustic neuromas, we measured the brainstempotentials. All of them were pathological, i.e. no false-negative results. The reduction of false-positive results depended upon the accuracy of the stimulation and recording techniques. 2 patients with suspected acoustic neuromas because of questionable cisternography findings had a normal latency between wave I and wave II. Surgery revealed and arachnoiditis scar in 1 case and a vessel loop of the anterior inferior cerebellar artery causing a block in the contrast medium in the other patient.

The review of the above-described otological and radiological data caused a change in our routine procedure for detecting acoustic neuromas. A complete 'testbattery' is no longer used. We now perform only a pure tone audiogram, a vestibular test including the observation of spontaneous and provocational nystagmus, a caloric test and a plain X-ray by Stenvers.

If two results in these groups (audiologic, vestibular, X-ray) are pathological, brain stem audiometry is performed. If an inner ear hearing loss progresses slowly over 6 months we also advise brain-stem evoked-response audiometry (BERA). If the results in BERA are pathological, a computerized tomography is performed. If a pathological process can already be seen with intravenous contrast medium, we have to assume extensive lesion. If this procedure does not show a pathological process, an air filling of the cerebello pontine cistern is performed by lumbal puncture.

If the air filling of the internal auditory meatus remains doubtful, the possibility of an exploratory function preserving transtemporal opening of the internal auditory meatus is discussed with the patient and strongly advised. BERA and computerized tomography including air filling of the posterior cisterns may be repeated once or twice a year to ensure a safe diagnosis. Therapy in all pathological cases is surgery which could be the topic of another meeting.

References

1   Fisch, U.; Wegmüller, A.: Early diagnosis of acoustic neuromas. ORL *36:* 129–140 (1974).
2   Grehn, S.; Helms, J.: Der erweiterte «leere» innere Gehörgang. Fortschr. Geb. Röntg Strahl Nukl. Med. *124:* 150 (1976).
3   Helms, J.: Zur Differentialdiagnose des Kleinhirnbrückenwinkeltumors. Lar. Rhinol *53:* 194 (1974).
4   Helms, J.; Grehn, S.: Atrophy of the labyrinthine nerves presenting with symptoms of acoustic neurome. Anais VI Simp. Ibero-Am. Otoneurolg., 159 (1974).
5   Koppenburg, P.: Leitsymptom Innenohrschwerhörigkeit. Befundstatistik; Diss. Tübingen (1978).

6   Maurer, K.; Schäfer, E.; Abdel Aziz, M.Y.; Leitner, H.: Veränderungen der akustisch evozierten Potentiale (FAEP) beim Akustikusneurinom (AN). Hörgeräte-Akust., suppl.,pp. 29–38 (1979).

7   Sortland, O.: Computertomography combined with gas-cisternography for the diagnosis of expanding lesions in the cerebellopontine angle. Neuroradiology *18:* 19–22 (1979).

8   Wende, S.: Die kombinierte Anwendung von Computer-Tomographie und Luftcisternographie bei Kleinhirnbrückenwinkel-Tumor. Fortschr. Geb. Röntg Strahl. Nukl. Med. *132:* 666–667 (1980).

Prof. Dr. med. J. Helms, Direktor der Universitäts-HNO-Klinik,
Langenbeckstrasse 1, D-6500 Mainz (FRG)

Adv. Oto-Rhino-Laryng., vol. 30, pp. 138–140 (Karger, Basel 1983)

Neurotological Diagnosis of Intracranial Lesions

Arvind Kumar, Nicholas Torok

Department of Otolaryngology, Head and Neck Surgery (Head: Dr. *Edward L. Applebaum)*, University of Illinois College of Medicine, University of Illinois Hospital and Eye and Ear Infirmary, Chicago, Ill., USA

Introduction

In the last 30 years a variety of vestibular and audiometric tests have been developed, the prime objective of which is to indicate the site of lesion responsible for the hearing loss or dizziness. In vestibular testing, the caloric test remains the most useful indicator of vestibular sensitivity and provides the means to assess the function of each side separately. However, a variety of caloric test techniques are being advocated and used in clinical practice. In addition, there are several types of rotational tests that use computer technology, optovestibular testing, quantitative analyses of eye movements, and tests of vestibulospinal function. The auditory brainstem response (ABR) is rapidly becoming an integral part of the neurotological armamentarium. The scope of conventional audiometric tests is also wide and increasing. No doubt some of these approaches have greatly increased our diagnostic capabilities but their nonselective use has created avoidable and often confusing redundance and contributed to increased cost.

To delineate the appropriate use of this expanding array of test procedures, involving the use of expensive technology and support personnel, it may be rewarding to assess the information yield of the more frequently used tests in neurotological practice. The purpose of this paper is to evaluate the reliability of the various tests routinely used at the University of Illinois.

Methods and Results

In this review 37 patients had complete vestibular evaluation, ABR and audiometric studies. The vestibular tests included (1) calibration, (2) eye tracking, (3) optokinetic tests, (4) Torok monothermal differential caloric test, (5) postural test, (6) search for spontaneous nystagmus, (7) Wodak past pointing test, (8) gait and Romberg in selected instances. All 37 patients exhibited one or more of the signs of retrolabyrinthine pathology [3].

The audiometric evaluation included (1) pure-tone air and bone audiogram, (2) speech reception and discrimination scores, (3) impedance studies, (4) site of lesion tests including tone decay, alternate binaural loudness balance test and Békésy audiometry, (5) central auditory tests including performance intensity function for phonetically balanced words, synthetic sentence identification with ipsilateral and contralateral competing messages, and staggered spondaic word test. The ABR testing was employed in all 37 patients. The criteria for an abnormal ABR were (1) interaural wave V latency difference equal to, or greater than 0.3 ms; (2) greater wave I-V interval than 2 standard deviations (SD) of normal; (3) greater wave V delay 2SD of normal; (4) lack of reproducibility of wave form with two averaged responses at the same intensity.

Of the 37 patients 7 did not complete the recommended confirmatory radiological studies such as computerized tomography (CT) scan and consequently no final diagnosis could be reached. All 30 remaining patients showed abnormal vestibular signs suggestive of posterior fossa pathology. In 26 patients (87%) a lesion in this area was confirmed. The pathological lesions detected were as follows: posterior fossa circulatory abnormalities (confirmed by CT-dynamic flow studies) in 8; acoustic neurinoma, 6; cerebellopontine angle meningioma, brainstem cerebellar cyst, and multiple sclerosis, each in 2; and cerebellar metastatic tumor, plexiform neurofibromatosis, basilar artery aneurysm, neurosyphilis, and prominent cerebellar atrophy, each in 1 patient. 1 patient had a loop of the anterior inferior cerebellar artery within the internal auditory canal. In 4 patients no lesions could be detected by radiology.

In the 26 patients where a pathological lesion was confirmed in the posterior fossa, ABR was abnormal in 14. It was reported as normal in 10. In 2 patients the test could not be done because of a profound hearing loss. In the 4 patients where no lesions were detected in the posterior fossa, ABR was abnormal in 3 and normal in 1.

A complete audiometric evaluation was performed in 7 patients as described above. In the remaining patients at least pure-tone audiogram with speech discrimination score was obtained with some having site of lesion tests as well. In 3 of the 7 patients where a comprehensive audiological assessment was made, the results did not contribute toward the final diagnosis. In these 3 patients the diagnosis confirmed by CT scan was acoustic neurinoma, basilar artery aneurysm, and a large dermoid cyst of the IVth ventricle.

Discussion

If the validity of a test is the consistency with which positive or negative findings agree with the actual presence or absence of disease, then the vestibular evaluation is both sensitive and consistent for detecting intracranial posterior fossa lesions. Utilizing techniques developed at the University of Illinois which include photoelectric nystag-mography [1] and the Torok monothermal differential caloric test [2], we have shown that central vestibular signs are 93% accurate in detect-ing posterior fossa lesions [3].

In the present series again, the vestibular evaluation has proved its value. The development of the CT scan has contributed immeasurably to the credibility of vestibular testing. Before the widespread use of the CT scan and before current technical refinements such as thin-section tomography, CT-regional blood brain studies and pneumo-CT were developed, the confirmation of a suspected lesion in the posterior fossa was limited. Though the patient often continued to be symptomatic little credence was given to the clearly abnormal vestibular test results. This is now no longer the case. As the results of the present study indi-cate a variety of pathological lesions in the posterior cranial fossa can be detected by a vestibular evaluation.

The question arises about the value and reliability of ABR, central auditory tests and impedance studies in neurotologic practice when the diagnostic yield from the outlined vestibular evaluation is so much more contributory. Do these tests yield information which significantly negates or enhances data already available from the vestibular evalua-tion? There is convincing evidence that when the vestibular evaluation indicates some pathology in the posterior fossa which could be con-firmed by advanced radiology, the audiologic studies and ABR do not add information concerning the pathological change and diagnosis in the intracranium.

References

1   Torok, N.; Guillemin, V.; Barnothy, J.M.: Photoelectric nystagmography. Ann. Otol. Rhinol. Lar. 60: 917–926 (1951).
2   Kumar, A.: The diagnostic advantages of the Torok monothermal differential cal-oric test. Laryngoscope 91: 1679–1694 (1981).
3   Kumar, A.; Torok, N.; Valvassori, G.E.: Central vestibular signs, posterior fossa pathology and computerized tomography-regional blood brain circulation. Ann. Otol. Rhinol. 90: 624–629 (1981).

A. Kumar, MD, FRCS, Department of Otolaryngology, Head and Neck Surgery, University of Illinois College of Medicine, and University of Illinois Hospital and Eye and Ear Infirmary, 1855 West Taylor Street, Chicago, IL 60612 (USA)

Adv. Oto-Rhino-Laryng., vol. 30, pp. 141–149 (Karger, Basel 1983)

Vestibular Hyperreactivity in Patients with Multiple Sclerosis

Patrick L.M. Huygen

ENT Department, University of Nijmegen, The Netherlands

Introduction

Vestibular hyperreactivity (VH) has been reported in a wide range of percentages (20–70%) of patients with multiple sclerosis. Prolonged duration, increased amplitude and frequency of vestibular nystagmic responses have been noted. Many reports concern caloric responses, but this report will be confined to rotatory responses. Bárány's test (10 revolutions in 20 s) was first used [*Ohm*, 1935; *Bentzen* et al., 1951; *Aubry and Pialoux*, 1957]. This test has the drawback that postrotatory responses may be influenced by persisting perrotatory effects. As an additional complication, nystagmic responses have been evaluated with the patient's eyes open in the light [*Ohm*, 1935, is one exception]. VH may then be imitated by lack of fixation suppression [*Takemori*, 1977]. Harmonic acceleration (HA) has been used [*Guerrier* et al., 1971; *Belhakia* et al., 1977; *Collard and Conraux*, 1980; *Sharpe* et al., 1981] and velocity steps (VS) have been applied by *von Albert* [1965]. A special type of hyperactive response was observed, which consisted of persistent secondary postrotatory nystagmus. This type and other types have been discussed by *Kornhuber* [1966]. Increased gain is one type and another type is the induction of alternating nystagmus, called 'multiple rebound' by *Hood* [1981], which also occurs spontaneously in multiple sclerosis [*Kornhuber*, 1959]. This study is devoted to VH in response to VS and HA stimulation. Correlations with and among various oculomotor abnormalities were also investigated.

Table I. Normal values for VS and HA tests obtained from 9 subjects (2 female, 7 male, 24–38 years of age) without any vestibular or hearing abnormalities

	VS (90°/s)	HA (0.05 Hz, 28°/s[a])
Gain	0.53 (0.34–0.72)[b]	0.44 (0.21–0.66)
Initial velocity, °/s	48 (31–65)	
Maximum SPV, °/s		12 (6–18)
Time constant, s	16 (9–24)	

[a] Maximum velocity.
[b] Mean (95% confidence limits).

Table II. Manifestations of VH in multiple sclerosis patients (number/total number)

	Normal	Too high		
		R	L	R and/or L nystagmus
VS				
Gain	0.72	28/71 (39%)	30/72 (42%)	34/74 (46%)
Time constant	24	9/69 (13%)	9/71 (13%)	15/72 (21%)
HA				
Gain	0.66	25/65 (38%)	21/65 (32%)	32/65 (49%)
VS and/or HA any parameter				51/78 (65%)

Material and Methods

Vestibular and oculomotor responses were evaluated in 80 multiple sclerosis patients, referred from the Neurology Department. Only oculomotor phenomena were evaluated in another 15 patients. A Tönnies equipment was used for evoking optokinetic nystagmus by projecting shadows moving over a half-cylindrical screen at 40, 60, and 80°/s, for rotatory stimulation (eyes open in darkness) with either VS (90°/s) or HA (0.05 Hz, 90° amplitude, 28°/s maximum velocity) and for DC electro-oculography (0–100 Hz) with several monocular and binocular leads for horizontal and vertical movements. Paper and analog recording was done with an Elema jet-ink writer and a Bell & Howell tape recorder. Fixation suppression was tested during HA by using a visual target attached to the chair. Smooth pursuit (SP) was tested with a target moving in a circle of 10° radius at 0.3 Hz, 20°/s maximum velocity [Umeda and Sakata, 1975]. Eye movements during SP, optokinetic nystagmus, fixation suppression and gaze positions were evaluated from the electronystagmographic record in a classical way. Abnormal SP was detected by catch-up saccades or disorganized response, abnormal optokinetic nystagmus by measuring slow phase velocity (SPV) values below 20°/s [*Oey*, 1981]. Defective fixation suppression was concluded if SPV values were higher than 0.5 of the values during HA in darkness [*Takemori*, 1977]. Vestibular nystagmus was analyzed with the method of nystagmometry [*Huygen*, 1979], using an off-line PDP 11/45 system, after A/D conversion at 250 Hz. Saccade velocity and acceleration profiles were calculated from monocular eye positions in 20° horizontal calibration signals with methods of numerical smoothing and differentiation. Saccadic slowing was indicated by maximum velocities below 300°/s at this amplitude [*Baloh* et al., 1975]. Gains for HA and VS were estimated from maximum and initial SPV values, respectively, in hard-copy graphs. SPV values for HA were plotted in periodical superposition and for the VS test on a logarithmic scale versus time. From the latter plot, time constant and initial velocity were obtained by visually fitting, inter- and extrapolating a straight line. VH was assumed if one or more response parameters were beyond the upper confidence limits established in 9 normal subjects (table I).

Results

Various types of VH were observed in high incidence (table II). Gains up to 1.4 and time constants up to 34 s were seen. High gain VH is shown in figure 1a and figure 1b illustrates VH with high time constant. In 2 patients with VH, alternating nystagmus was induced and in 2 others secondary postrotatory nystagmus persisted in both directions. In VS responses gain and time constant showed negative correlation. High gain in HA (32 cases) was most often associated with high gain in the VS test for one or both nystagmus directions (17 cases). Association of high HA gain with high time constant was also found (6 cases). In 5 cases with significantly high gain in HA, no VH was seen in the VS test, though both time constant and gain were high enough to account for a synergistic effect.

Among the 95 patients in whom oculomotor functions were examined, 37 had saccadic slowing. 24 patients had bilateral nasal slowing (INO), in 3 of them combined with unilateral temporal slowing (cf. one-and-one-half syndrome). 5 patients had unilateral INO. 1 patient had isolated temporal slowing (VI paresis) on both sides and 3 had it on one side. Bidirectional slowing was seen in 1 patient in one eye and in 3 patients in both eyes. Defective SP, always in combination with optokinetic nystagmus and fixation suppression deficit, was found in 35 patients and was significantly correlated with saccadic slowing: the combined abnormalities were seen in 22 (out of 95) patients. Among the 80 patients fully examined, VH by whatever parameter (table II)

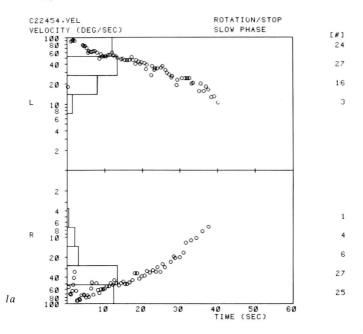

1a

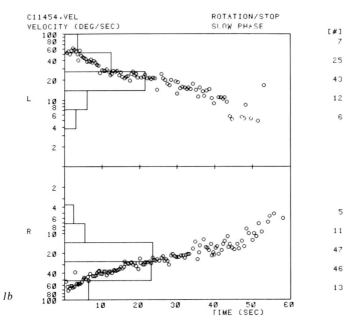

1b

C38801.DAT   DETAILED SACCADE ANALYSIS

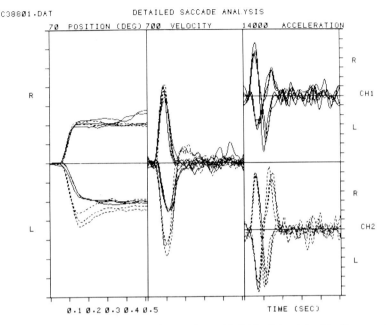

Fig. 2. Detailed saccade analysis in a patient with INO of the right side. Superposition of saccades of 20° amplitude for right eye (continuous line) and left eye (dashed line) in position and velocity plots. Superposition of saccades in acceleration plot for right and left eye, separately. —— = CH1; – – – = CH2.

was not correlated with SP deficit (26/80), saccadic slowing (29/80) or clinical signs of cerebellar involvement (40/80).

A factor complicating analysis of nystagmic responses was recognized in cases with specific oculomotor disturbances involving gaze nystagmus. This is exemplified by a patient having INO of the right side. Saccadic slowing of the right eye to the left is shown in figure 2. Abducting nystagmus of the left eye was seen on gaze to the left. Exaggeration of the HA response is to be noted in the response of the left eye while abducting (fig. 3a, second trace), whereas the response of the right eye (top trace) is symmetric. SPV values calculated from the bin-

Fig. 1. Hyperactive VS responses in 2 different patients. *a* Gain too high (initial velocity over 100 for right and 80 for left beating nystagmus). Note insufficient damping (time constant 29 s) for left beating nystagmus. *b* Time constant too high (32 s for right, 27 s for left beating nystagmus).

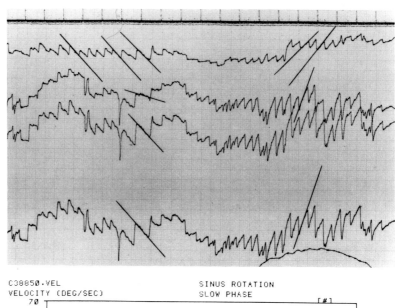

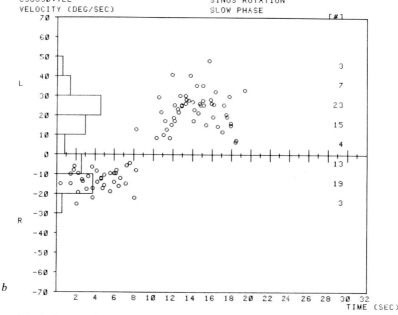

Fig. 3. Same patient as in figure 2, HA response. *a* electronystagmographic record with response of right eye (top trace), left eye (second trace) and summated signal (third and bottom traces). *b* Computer display from summated signal. SPV values plotted in periodical superposition. Histograms are included for facilitating the detection of asymmetry.

ocular signal also show the disturbance (fig. 3b). Similar findings have been reported by *Baloh* et al. [1978]. In the primary VS response the logarithmic plot of SPV versus time usually is rectilinear except for the final part, which shows transition to the secondary response phase. The disturbing component of gaze nystagmus is recognized as an upward thrust of the response curve over the first part of the response. In most cases, discard of that part of the plot allowed for a reasonable fit to the remaining SPV values.

Discussion

As early as 1935, *Ohm* suggested a common cause of optokinetic nystagmus deficit and VH. *Sharpe* et al. [1981] adopted the suggestion made before in relation to cerebellar degeneration by *Zee* et al. [1976] that VH may indicate plastic adaptation of the vestibulo-ocular reflex to low-gain SP. It is tempting to speculate on cerebellar dysfunctioning as a cause of VH, especially if combined with SP and fixation suppression deficits [*Dichgans and Jung,* 1975 for review], but there was no significant correlation of VH with either SP deficit or cerebellar involvement in the present series. The latter correlation was also lacking in the series of *Sharpe* et al. [1981], but *Salonna and Carbonara* [1959], on the other hand, emphasized the correlation between VH and cerebellar deficit. An important correlation was seen between SP deficit and saccadic slowing. This fits in with the report by *Mastaglia* et al. [1979]. Perhaps, it can be inferred that the brain stem was more involved in SP deficit than the cerebellum, since no correlation was found between cerebellar involvement and SP deficit.

In some multiple sclerosis patients with VH, feelings of 'sickness' occurred after vestibular stimulation, which is reminiscent of motion sickness. The clinical impression is that motion sickness does occur in some patients in relation to their disease. In particular cases, motion sickness was even used as an argument against rotatory testing. More typical, however, as has been reported by others [*Aubry and Pialoux,* 1957; *Salonna and Carbonara,* 1959], may be the fact that patients with VH can be strikingly insensitive to vestibular stimulation.

It may be interpreted as a manifestation of vestibulo-ocular reflex plasticity that gain and time constant exhibited a negative correlation. Interestingly, there is a parallel with *Groen's* [1965] observation that during postnatal development in the dog both time constant and threshold decreased. His interpretation was in favor of 'central inhibition' coming into action. In this view, a high time constant would indicate release from central inhibition. Release from cortical inhibition, to be more specific, might explain the finding reported by *Calliauw* [1968] that 4 out of 5 hemispherectomy patients developed VH after surgery with very high time constants. It is worth of note that high time constants have also been found in seasickness [*van Egmond* et al., 1954] and airsickness [*Aschan*, 1954; *Krijger*, 1954]. In fighter pilots the time constant appeared to be inversely related to the amount of (recent) flight experience. The (short) time constant became 'normalized' in a pilot who was off duty for several months. This phenomenon can be interpreted as indicating vestibular dishabituation. Such conception may be applicable to mutiple sclerosis, which leads to the suggestion that the time constant becomes higher as the patient becomes more and more immobilized.

Acknowledgement

I am indebted to Prof. *O.R. Hommes* for referring the patients and to Mr. *M.G.M. Nicolasen* for assistance in patient examination and computer analysis.

References

Albert, H.-H. von: Elektronystagmographische Untersuchungen über die Trainierbarkeit der Labyrinthe, bzw. der Vestibulariskomplexe, gemessen am postrotatorischen Nystagmus I und II. Arch. Ohr.-Nas.-KehlkHeilk. *173:* 371–383 (1965).

Aschan, G.: Response to rotatory stimuli in fighter pilots. Acta oto-lar., suppl. 116, pp. 24–31 (1954).

Aubry, M.; Pialoux, P.: Maladies de l'oreille interne et oto-neurologie (Masson, Paris 1957).

Baloh, R.W.; Konrad, H.R.; Sills, A.W.; Honrubia, V.: The saccade velocity test. Neurology *25:* 1071–1076 (1975).

Baloh, R.W.; Yee, R.D.; Honrubia, V.: Internuclear ophthalmoplegia. II. Pursuit, optokinetic nystagmus, and vestibulo-ocular reflex. Archs Neurol. *35:* 490–493 (1978).

Belhakia, A.; Ben Hamida, M.; Bouzouita, H.: Sclérose en plaques en Tunisie: étude cochléo-vestibulaire. Annls Oto-lar., Paris *94:* 617–622 (1977).

Bentzen, O.; Jelnes, K.; Thygesen, P.: Acoustic and vestibular function in multiple sclerosis. Acta psychiat. neurol. scand. *26:* 265–295 (1951).

Calliauw, L.J.E.: Akoestische, vestibulaire en optokinetische funkties na hemispherectomie bij de mens (English summary); thesis Utrecht (1968).

Collard, M.; Conraux, C.: L'électronystagmographie au cours de la sclérose en plaques. Annls Oto-lar., Paris 97: 467–482 (1980).

Dichgans, J.; Jung,. R.: Oculomotor abnormalities due to cerebellar lesions; in Lennerstrand, Bach-y-Rita, Basic mechanisms of ocular motility and their clinical implications, pp. 281–298 (Pergamon Press, Oxford 1975).

Egmond, A.A.J. van; Groen, J.J.; Wit, G. de: The selection of motion sickness-susceptible individuals. Int. Rec. med. 167: 651–660 (1954).

Groen, J.J.: Central regulation of the vestibular system. Acta oto-lar. 59: 211–218 (1965).

Guerrier, Y.; Bassères, F.; Mouchard, J.: Les aspects électronystagmographiques de la sclérose en plaques. Cah. ORL 6: 348–367 (1971).

Hood, J.D.: Further observations on the phenomenon of rebound nystagmus. Ann. N.Y. Acad. Sci. 374: 532–539 (1981).

Huygen, P.L.M.: Nystagmometry: the art of measuring nystagmus parameters by digital signal processing. ORL 41: 206–220 (1979).

Kornhuber, H.H.: Der periodisch alternierende Nystagmus (Nystagmus alternans) und die Enthemmung des vestibulären Systems. Arch. Ohr.-Nas.-KehlkHeilk. 174: 182–209 (1959).

Kornhuber, H.H.: Physiologie und Klinik des zentralvestibulären Systems (Blick und Stützmotorik); in: Berendes, Link, Zöllner, Hals-Nasen-Ohrenheilkunde, vol. III, pp. 2150–2351 (Thieme, Stuttgart 1966).

Krijger, M.W.W.: De betekenis van het evenwichtsorgaan voor de vlieger (English summary) thesis Utrecht (1954).

Mastaglia, F.L.; Black, J.L.; Collins, D.W.K.: Quantitative studies of saccadic and pursuit eye movements in multiple sclerosis. Brain 102: 817–834 (1979).

Oey, P.L.: Enige aspecten van het optokinetische onderzoek (English summary); thesis Amsterdam (1981).

Ohm, J.: Über die Beziehungen zwischen vestibulärem, willkürlichem und optokinetischem Nystagmus bei der multiplen Sklerose. Dt. Z. NervHeilk. 138: 43–64 (1935).

Salonna, F.; Carbonara, L.: Reperti otovestibolari in soggetti affetti da sclerosi a placche. Oto-rino-lar. ital. 28: 233–241 (1959).

Sharpe, J.A.; Goldberg, H.J.; Lo, A.W.; Herishanu, Y.O.: Visual-vestibular interaction in multiple sclerosis. Neurology 31: 427–433 (1981).

Takemori, S.: Visual suppression test. Ann. otol. 86: 80–85 (1977).

Umeda, Y.; Sakata, E.: The circular eye-tracking test. ORL 37: 290–298 (1975).

Zee, D.S.; Yee., R.D.; Cogan, D.G.; Robinson, D.A.; Engel, W.K.: Ocular motor abnormalities in hereditary cerebellar ataxia. Brain 99: 207–234 (1976).

P.L.M. Huygen, PhD, ENT Department, University of Nijmegen,
Philips van Leydenlaan 15, NL-6500-HB Nijmegen (The Netherlands)

Adv. Oto-Rhino-Laryng., vol. 30, pp. 150–155 (Karger, Basel 1983)

'Recruitment/Decruitment' of Caloric Responses: Effect of Anatomical Variables

L.R. Proctor[a], *R.K. Frazer*[b], *R.N. Glackin*[a], *C.R. Smith*[c]

[a] Department of Laryngology and Otology, [b] Applied Physics Laboratory, and
[c] Department of Medicine, The Johns Hopkins University, Baltimore, Md., USA

Several authors have reported using multiple incremental caloric stimulations to study the input/output (I/O) relationship of the vestibuloocular reflex in disease states [1–3]. For example, *Torok* [2] reported that a flattened I/O profile, which he termed 'vestibular decruitment', was characteristic of central, rather than peripheral, vestibular disorders. We found, among normal subjects, a consistent variation in the I/O profile, which we believe is related to variations in temporal bone structure [4, 5].

A mathematical model can be used to predict the time course of the temperature difference (ΔT) across the lateral semicircular canal, based on temperature events within the aural canal [6]. For simplicity, the value of ΔT is equated with the magnitude of the caloric stimulus. By switching the temperature of a continuous aural irrigation from hot to cold values at precisely specified times, the model indicates that the magnitude of ΔT can be adjusted to a desired peak level (ΔT^*) and then promptly eliminated (fig. 1). We applied seven caloric stimulations of the type shown in figure 1 to 30 healthy subjects on 3 successive days, along with impulsive rotatory tests [7]. The first four caloric stimulations simulated the conventional test battery. After a rest period, a series of three incremented stimulations was applied (I/O tests). A somewhat stronger set of stimulations (table I) was required for 10 subjects (group A) than was used for the remaining 20 subjects (group B). Nystagmus responses to rotatory and caloric stimulations were analyzed on a digital computer [7].

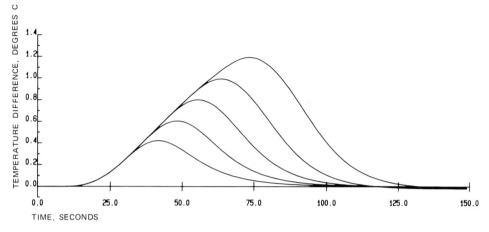

Fig. 1. Predicted temperature difference across lateral semicircular canal (ΔT) resulting from hot/cold aural irrigation pulses of various durations.

Table I. Caloric stimulation levels (expressed as ΔT^*) applied to 30 subjects

	Basic test battery	Input/output tests		
		low	medium	high
Group A (n = 10)	1.0	0.6	0.8	1.2
Group B (n = 20)	0.8	0.4	0.6	1.0

Table II. Correlation between I/O scores and individual I/O tests, two 'basic battery' tests (hot, left; cold, left) and postrotatory test (Pearson's r statistic)

	Low I/O	Medium I/O	High I/O	Hot	Cold	Post-rotatory
Group A (n = 10)	−0.7181 (0.010)	0.1383 (0.352)	0.8744 (<0.001)	0.2248 (0.266)	0.2953 (0.204)	−0.5517 (0.049)
Group B (n = 20)	0.4882 (0.017)	0.8045 (<0.001)	0.9548 (<0.001)	0.9256 (<0.001)	0.8345 (<0.001)	0.2006 (0.205)
All cases (n = 30)	−0.0028 (0.494)	0.6249 (<0.001)	0.9203 (<0.001)	0.7865 (<0.001)	0.5713 (0.001)	−0.1157 (0.275)

Probability statistic in parentheses.

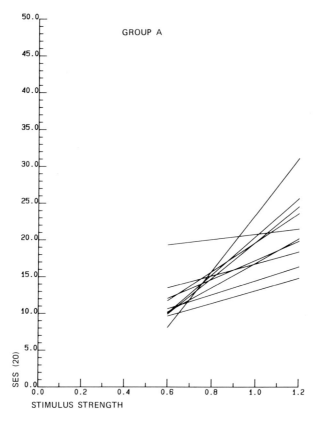

Fig. 2. Linear regression (I/O slope) lines fitted to averaged response scores for group A subjects. SES (20), slow phase eye speed averaged over maximum 20-second segment of response. Stimulations (ΔT^*) were at 0.6, 0.8, and 1.0 °C.

I/O test scores did not change significantly during 3 successive days. A regression line was computed (I/O slope) on averaged data from the 3 days for each subject. As seen in figures 2 and 3, group B had steeper I/O slopes ($\bar{x} = 25.47$) than did group A ($\bar{x} = 16.83$, $p < 0.05$). The difference in I/O slope values could result from a physiologic tendency to hyperresponsiveness in group B vs. group A; such hyperresponsive subjects also might produce a steeper I/O profile. If this were true, responses to rotation should have been correlated with I/O slope. Instead, no significant positive correlation was found (table II). In contrast, note that most caloric test scores (table II) were

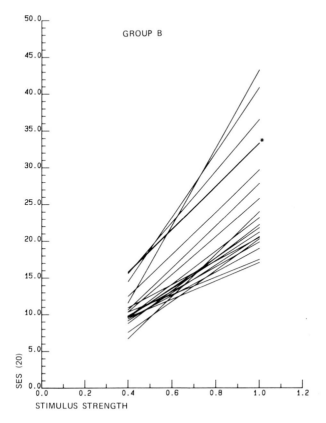

Fig. 3. I/O slope lines for group B subjects. Asterisk indicates two superimposed slopes. Stimulations were at 0.4, 0.6, and 1.0 °C.

correlated significantly with I/O slope for the group as a whole as well as for group B considered individually. The results for group A are of uncertain implication, in view of the small number of cases.

The mathematical model indicates how variations in the distance between the aural canal and the center of the lateral semicircular canal (x, fig. 4) could explain a positive relationship between I/O slope and overall responsiveness to caloric stimulation. Figure 4 shows how the model relates the stimulus magnitude (fig. 1) to the duration of the primary (hot) irrigation pulse for varicus values of x. Note that for smaller values of x there are predicted both a stronger stimulation and a steeper

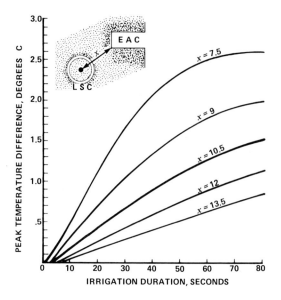

Fig. 4. Peak temperature differences (ΔT^*) across the lateral semicircular canal predicted to result from various irrigation pulse durations at various distances (x) between the lateral semicircular canal (LSC) and the aural canal surface (EAC). Assumed typical distance: x = 10.5 mm.

growth of the caloric stimulus, while for larger values of x there are predicted both weaker stimulations and a flatter growth of stimulus. These results support the concept that the steepness of the I/O profile of caloric responses in normal individuals may be determined in large part by anatomical characteristics of the temporal bone in the labyrinth area. This problem needs further study, to establish a rational basis for assessing vestibular function through caloric stimulations.

Acknowledgements

We thank Prof. A. Kimball, Department of Biostatistics, for his help in the preparation of this report. This work was supported in part by USPHS Grant CMS 5 R01 13780–04, by the Department of the Navy N0024-81–C–5301 and by a grant from the Schering Corporation.

References

1 Litton, W.B.; McCabe, B.F.: Neural vs. sensory lesion: vestibular signs. Laryngo-scope, St. Louis *76:* 1113–1127 (1966).
2 Torok, N.: A new parameter of vestibular sensitivity. Ann. Otol. Rhinol. Lar. *79:* 808–817 (1970).
3 Ghosh, P.; Kacker, S.K.: Vestibular recruitment and decruitment. Acta oto-lar. *88:* 227–234 (1979).
4 Zangemeister, W.H.; Bock, O.: The influence of pneumatization of mastoid bone on caloric respones. A clinical study and a mathematical model. Acta oto-lar. *88:* 105–109 (1979).
5 Proctor, L.R.: The effect of variations in temporal bone structure upon the caloric response. Acta oto-lar. *94:* 253–259 (1982).
6 Proctor, L.R.: A new approach to caloric stimulation of the vestibular receptor. Ann. Otol. Rhinol. Lar. *84:* 683–695 (1975).
7 Proctor, L.R.; Glackin, R.N.; Smith, C.R.; Lietman, P.S.: A test battery for detecting vestibular toxicity; in Lerner, Matz Hawkins, Aminoglycoside ototoxicity. (Little, Brown, Boston 1981).

L.R. Proctor, MD, Department of Laryngology and Otology,
The Johns Hopkins University School of Medicine, Baltimore, MD 21205 (USA)

Adv. Oto-Rhino-Laryng., vol. 30, pp. 156–158 (Karger, Basel 1983)

Ocular Fixation Test during Damped Pendular Rotation Test and Auditory Brain Stem Response for the Differential Diagnosis of Central Vestibular Disorders

Jin Kanzaki, Toshiaki O-Uchi

Department of Otolaryngology, School of Medicine, Keio University, Tokyo, Japan

Recent physiological research on the vestibuloocular reflex has shown that the function of visual or ocular fixation has some relation to the cerebellar flocculus and nodulus, and that cerebelloflocculectomy causes disorders of ocular fixation. These results are now in use for clinical purposes.

We have reported previously on our investigations of central vestibular disorders by the damped pendular rotation test (DPRT) [1] and further, on some part of our study of the ocular fixation test (OFT) by means of DPRT [2]. The aim of this study is to evaluate the diagnostic value of using OFT by DPRT and DPRT in the dark with eyes open jointly with auditory brain stem response (ABR) audiometry as a battery test for the differential diagnosis of central vestibular disorders.

72 cases of central disorders composed the main part of the present study. Disorders revealed by OFT were classified into the following four grades:

++ = Perrotatory nystagmus, same as that found by DPRT given in the dark with eyes open, evoked in both directions or in one direction.

+ = Perrotatory nystagmus evoked, but with less frequency and small amplitude.

± = No evoking of nystagmus, however, obvious irregular ocular movement.

– = Neither nystagmus nor ocular movement.

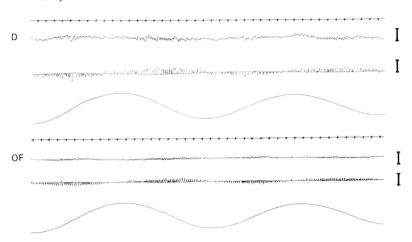

D = Recordings in the dark with eyes open; OF = recordings under ocular fixation.

Fig. 1. OFT in a case of ependymoma of the IVth ventricle. D = Recordings in the dark with eyes open; OF = recordings under ocular fixation. The nystagmic rhythm in D is regular. Under ocular fixation, no suppression of nystagmus is revealed. Calibration 20°/s.

Results

Brain Stem Lesions

In cases of tumor of the lower brain stem, there was no perrotatory nystagmus evoked by DPRT given in the dark with eyes open; however, contrary to this finding, nystagmus was evoked by OFT. By ABR audiometry, abnormality was in most cases obvious on both sides and disappearance of the wave pattern and/or a prolonged latency were noted.

In upper brain stem lesions, findings of DPRT in the dark were normal or had only a slight degree of disorder; however, under ocular fixation, the disorder was found to be grade + +. ABR audiometry was normal. In IVth ventricle tumors, the nystagmic rhythm was regular by DPRT in the dark, but under ocular fixation, the disorder was classified as grade + + (fig. 1).

By ABR, prolongation of the interpeak latency was found between III and IV as well as between IV and V and prolonged latency after wave III.

Cerebellar Lesions

In vermis tumors or in the cases of degeneration or atrophy, the disorders were determined as grade + or ++ under ocular fixation. ABR was normal. With respect to tumors in the hemisphere, dysrhythmia which could be a secondary effect of the tumor on the brain stem was sometimes indicated by DPRT given in the dark and, under ocular fixation, the disorder was classified as grade +. By ABR, a prolonged latency of wave V, which could also be a secondary effect on the brain stem, was recognized.

Acoustic Neuroma

Under ocular fixation 6 out of 29 cases (20%) of acoustic neuroma indicated disorders of grade + or ++. In observing the relation between the grade of disorder by OFT and ABR findings, it was discovered that many of the patients who had disorders of grade + or ++ had large tumors and were totally deaf, and so ABR audiometry could not be applied. In 9 out of the 17 cases which indicated grade ± or –, a prolonged latency was found in wave V, and 3 cases showed only wave I.

Even with these test findings, the differential diagnosis between upper brain stem lesions and cerebellar vermis lesions seemed to be difficult without neurological findings and CT scans. However, in the differential diagnosis of central vestibular disorders other than the lesions mentioned above, the use of OFT by DPRT in combination with ABR was quite effective.

References

1   Kanzaki, J.; Sakagami, C.: The damped pendular rotation test in central vestibular disorders. Archs Oto-Rhino-Lar. *214:* 97–107 (1976).
2   Kanzaki, J.: Effects of ocular fixation on perrotatory nystagmus in damped pendular rotation test. Archs Oto-Rhino-Lar. *230:* 209–219 (1981).

J. Kanzaki, MD, Department of Otolaryngology, School of Medicine, Keio University, Tokyo (Japan)

Adv. Oto-Rhino-Laryng., vol. 30, pp. 159–164 (Karger, Basel 1983)

Diagnosis of Peripheral and Central Vestibular Lesions by the Harmonic Acceleration Test

R. Probst, M. Aoyagi, C.R. Pfaltz

Department of Oto-Rhino-Laryngology, University of Basel, Switzerland

Introduction

The present study was carried out in order to test the diagnostic importance of the sinusoidal harmonic acceleration test. Until now substantial information about its diagnostic range of application is rather scarce [1–4]. The question still remains open whether the three parameters – phase lag, gain and directional preponderance – allow a reliable differentiation between peripheral and central vestibular lesions, one of the major problems in neurootology.

Methods and Materials

We used a computerized rotary chair of the Contraves-Goertz corporation with a controlled torque motor providing an accuracy of 0.05°/s. The test was run with a sinusoidal stimulus of five frequencies from 0.01 to 0.16 Hz in a soundproof and completely dark room. The subjects were kept alert by simple arithmetic tasks. Eye movements were recorded by ENG technique and processed by an on-line computer. The fast phases of nystagmus were clipped to zero and the slow phases averaged to one cycle. The test was run over four cycles for the lowest frequency and over eight cycles for all other frequencies. Peak velocities were held constant at 50°/s. The acceleration varied from 3 to 50°/s^2. For each frequency the phase difference relative to the chair acceleration (not to the velocity as often used), the gain and the preponderance were computed [5–6]. We tested 94 patients who presented with some pathologic vestibular findings such as spontaneous nystagmus, positional nystagmus or abnormal caloric responses. There were 49 with peripheral and 34 with central vestibular disorders, and 10 with Menière's disease. 1 patient was not classifiable. Our control group includes 86 persons with no history of vestibular

Table I. Pathologic test results: all patients (n = 94)

	Condition A		Condition B	
	n	%	n	%
All parameters	49	52	59	63
Phase lag	37	39.5	42	44.5
Preponderance	32	34	31	33
Gain	15	16	26	27.5

Condition A: value at one frequency more than 2 SD out of normal range; condition B: value at three or more frequencies 1 SD out of normal range.

disorders, ranging in age from 18 to 84 years. We found a nearly linear response for the *phase lag* relative to the chair acceleration. *Preponderance* shows slightly positive values. We also evaluated the *gain* of normal persons and patients. The range of normal distribution is rather wide and there is no correlation between individual gains versus age for all frequencies. The slow rise of the mean gain with increasing frequency may indicate the tested frequencies to be too high for a linear response. An effect of adaption to darkness may also be considered, especially in the higher frequencies.

Results

We used two conditions to evaluate pathologic results: *condition A* defines a test as pathologic if a value at one given frequency is exceeding 2 SD, i.e. the normal range of distribution. In *condition B* a test was considered pathological if the values of at least three of five frequencies were exceeding 1 SD. Using all three parameters (phase lag, gain and preponderance) together, only little more than 50% of the patients showed a pathologic test result, notwithstanding the evaluation method. *Phase lag* proved to be the most important parameter with positive findings in about 40%. *Gain* was the least sensitive parameter with only 16 or 27% pathologic results (Table I).

Patients with Ménière's disease (unilateral) showed quite consistently pathologic results (9 of 10 subjects). Phase lag was again the most sensitive parameter, gain and preponderance were both not very instructive. The graphic display of phase lag and gain seems to be more characteristic in the Ménière group than in others. The phase lag curve takes a steeper course than the template with a crossing point at 0.08

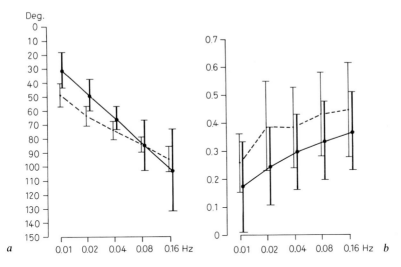

Fig. 1. Phase lag (a) and gain (b) from patients with Menière's disease (n = 10) as a function of frequency. Error bars indicate ± 1 SD. — = Normal subjects (n = 86).

Table II. Pathologic test results: central and peripheral groups

	Central group (n = 34)				Peripheral group (n = 49)			
	condition A		condition B		condition A		condition B	
	n	%	n	%	n	%	n	%
All parameters	15	44	22	64.5	24	49	28	57
Phase lag	12	35.5	17	50	17	34.5	18	36.5
Preponderance	10	29.5	12	35.5	18	36.5	15	30.5
Gain	4	12	9	26.5	8	16.5	13	26.5

Condition A: value at one frequency more than 2 SD out of normal range; condition B: values at three or more frequencies 1 SD out of normal range.

Hz as it was found by other authors [1, 2]. This seems to be a rather typical finding in Menière's disease. The gain does not show the same pattern. It is decreased in all frequencies and shows no crossover. Neither gain nor preponderance seem to provide any reliable diagnostic information (Fig. 1a, b).

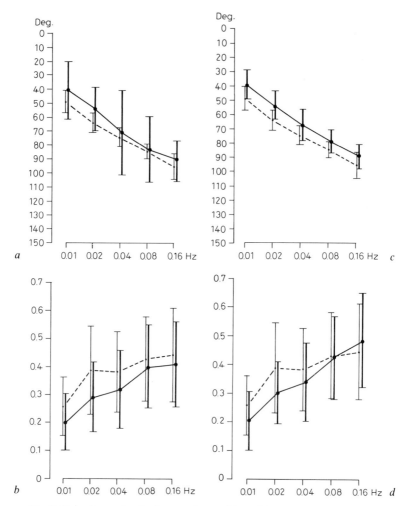

Fig. 2. Phase lag and gain from patients with peripheral vestibular lesions *(a, b)* (n = 49) and central vestibular lesions *(c, d)* (n = 34). Error bars indicate ± 1 SD. --- = Normal subjects (n = 86).

A comparison between the groups of peripheral and central vestibular lesions did not reveal any statistically significant differences. The sensitivity of the parameters was approximately the same. Again phase lag was the most reliable and gain the least informative parameter. Hence it may be assumed that peripheral and central lesions do not show any different pattern of parameter distribution. In the central group there is a clear shift of the phase lag curve to the right with relatively minor variations within this group. The peripheral group showed much greater variations. There is some tendency to a reduced phase lag in the lower frequencies but no specific pattern can be found. The gain within the two groups is more or less reflecting the findings of the phase lag but shows again a much wider variation than the former parameter. Preponderance proved to be pathologic in 30–35% in both groups (Table II, fig. 2).

Conclusions

The sinusoidal harmonic acceleration test seems to be a valuable and sensitive method in detecting vestibular dysfunction in Menière's disease. Phase lag is the most reliable parameter but the sensitivity of the test is better if all three parameters are used. For other vestibular disorders, there is an overall diagnostic sensitivity of only about 50%. Although a clear shift of the phase lag to lower values within the whole frequency range may indicate a central vestibular dysfunction, our results do not suggest that the sinusoidal harmonic acceleration test allows an accurate distinction between central and peripheral vestibular disorders. Moreover, the measurement of the gain by this method does not seem to be of any clinical use. However, there might be some other important diagnostic parameters or response patterns provided by the sinusoidal harmonic acceleration test, which we did not consider in the present study. More research is required to determine the value of this method as a diagnostic and prognostic test.

References

1   Wolfe, J.W.; Engelken, E.J.; Olson, J.W.; Kos, C.M.: Vestibular responses to bithermal caloric and harmonic acceleration. Ann. Otol. Rhinol. Lar. 87: 861–867 (1978).
2   Olson, J.E.; Wolfe, J.W.: Comparison of subjective symptomatology and responses to harmonic acceleration in patients with Menière's disease. Ann. Otol. Rhinol. Lar. 90: suppl. 86, pp. 15–17 (1981).
3   Rubin, W.: Sinusoidal harmonic acceleration test in clinical practice. Ann. Otol. Rhinol. Lar. 90: suppl. 86, pp. 26–28 (1981).

4 Simpson, R.A.: Sinusoidal harmonic acceleration labyrinthine test: clinical experience. Ann. Otol. Rhinol. Lar. *90:* suppl. 86, pp. 26–28 (1981).
5 Engelken, J.E.; Wolfe, J.W.: Analog processing of vestibular nystagmus for on-line
   cross-correlation data analysis. Aviat. Space envir. Med. *48:* 210–214 (1977).
6 Wolfe, J.W.; Engelken, E.J.; Kos, C.M.: Low-frequency harmonic acceleration as a
   test of labyrinthine function: basic methods and illustrative cases. Trans. Am. Acad.
   Ophthal. Oto-lar. *86:* 130–142 (1978).

R. Probst, MD, Department of Oto-Rhino-Laryngology, University Hospital,
CH-4031 Basel (Switzerland)

Adv. Oto-Rhino-Laryng., vol. 30, pp. 165–170 (Karger, Basel 1983)

Reducing the
Simple Harmonic Acceleration Test Time

M.A. Hamid

Department of Otolaryngology and Communicative Disorders, Cleveland Clinic
Foundation, Cleveland, Ohio, USA

Introduction

Sinusoidal harmonic acceleration (SHA) testing is a relatively new
and useful addition to the evaluation of the human vestibular function.
SHA offers controlled stimuli conditions, accurate response evaluation and increased sensitivity and repeatability compared to conventional thermal testing.

Presently, one of the commonly used systems (Contraves-Goerz,
DP-300) has three disadvantages: long test time (45 min), less emphasis
on the gain measure (eye velocity/chair velocity), and the high cost of
equipment.

The SHA test procedure comprises five trials at five harmonic frequencies: 0.01, 0.02, 0.04, 0.08, and 0.16 Hz. Each trial is composed of
eight cycles except for the 0.01 Hz, which uses four cycles only. The
number of cycles per trial, the processing technique and the normal
values of the phase lag and preponderance are adopted from *Wolfe* et
al. [1].

The rotational response is analyzed in three steps: (1) the quick
phase components are removed and a cumulative slow curve is formed,
(2) coherent averaging of the number of cycles over one complete cycle, and (3) the coherent average cumulative slow phase velocity (output) is cross-correlated with the chair velocity (input) to give estimates
of the phase shift, preponderance and gain measures.

Although rotational testing is a useful addition to our vestibular
examination battery, we were confronted with the problem of adding

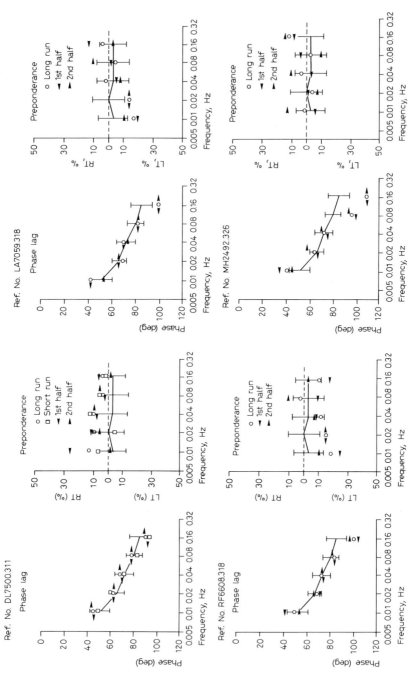

Fig. 1. Results of processing long-run and short-run of 4 control subjects.

Table I. Normal population: phase (latency)

Subject	Frequency (1st and 2nd half)									
	0.01 Hz		0.02 Hz		0.04 Hz		0.08 Hz		0.16 Hz	
1	−45.8	−42.7	−61.8	−61.8	−69.8	−66.7	−77.4	−77.4	−91.0	−87.3
2	−45.2	−47.7	−58.3	−59.5	−67.1	−67.4	−80.5	−80.2	−90.6	−90.7
3	−41.3	−27.2	−59.4	−59.6	−69.2	−67.2	−78.6	−82.7	−108.1	−104.5
4	−41.2	−52.4	−69.9	−65.6	−72.8	−72.1	−83.5	−81.3	−104.4	−97.5
5	−41.1	−46.8	−65.2	−72.13	−70.2	−67.7	−82.1	−82.5	−98.9	−98.0
6	−35.5	−43.5	−60.7	−65.1	−71.0	−67.0	−80.4	−83.1	−93.8	−96.2
7	−50.8	−39.4	−58.5	−61.3	−67.7	−67.7	−79.5	−74.0	−90.7	−93.8
8	−33.8	−43.8	−65.6	−58.5	−73.4	−69.8	−97.6	−94.8	−109.3	−109.9
9	−70.9	−41.7	−66.7	−63.4	−68.9	−63.8	−77.1	−77.3	−92.7	−93.7
10	−47.2	−47.2	−71.7	−70.9	−70.1	−73.0	−82.7	−85.4	−93.1	−92.4
Mean	−45.3	−43.2	−63.8	−63.8	−70.0	−68.2	−81.9	−81.8	−97.3	−96.4
SD	10.4	6.7	4.7	4.7	2.0	2.7	5.9	5.6	7.4	6.6
SCC	−0.93		0.93		0.43		0.65		0.86	
WX	19		27		8		26		21	

SD = Standard deviation; SCC = Spearman's coefficient of rank correlation; WX = Wilcoxon's sum of positive or negative ranks. 5% significance < 6. Nonparametric tests are used because of the relatively small number of observations and the underlying population is not normal.

an extra hour to each patient test time. Two possible solutions were considered: (1) reducing the number of trials, or (2) reducing the number of cycles per trial. To gain as much frequency response information as possible, the latter solution was adopted. In the following discussion, the term 'long-run' will mean eight cycles whereas 'short-run' will mean four cycles.

The purpose of this study is to show that the test time can be reduced to half without loss of significant information ($p < 0.05$).

Results and Discussion

10 normal adult control subjects underwent long SHA testing. Subject DL7500 underwent both long and short testing procedures. Each long test response was divided into two equal halves and each

Table II. Normal population: preponderance

Subject	Frequency (1st and 2nd half)									
	0.01 Hz		0.02 Hz		0.04 Hz		0.08 Hz		0.16 Hz	
1	27.0	1.7	11.4	6.2	8.3	11.8	2.1	5.5	6.3	-1.1
2	-26.1	14.8	-4.5	-2.2	3.7	1.6	1.6	-2.2	-1.7	0.7
3	7.5	14.8	-0.4	-2.2	1.4	3.1	4.3	-5.3	-0.2	-4.4
4	-24.5	-10.4	-13.8	-14.3	-8.8	-7.8	-8.6	10.5	-17.0	-2.5
5	-19.8	-11.6	-14.0	-15.1	-4.9	8.3	-1.9	-7.0	13.4	-0.6
6	5.5	2.2	-1.2	-6.5	1.9	-1.4	5.4	3.9	0.11	14
7	8.2	11.9	0.25	7.8	8.9	-5.3	-3.8	11.9	6.0	-5.9
8	-5.5	13.3	-1.2	-7.0	-2.8	11.1	3.9	-9.7	8.4	13.1
9	12.6	-11.9	-13.0	-1.8	-13.8	-12.7	-2.5	-5.8	-9.9	-14.5
10	1.0	1.0	-8.3	-10.4	5.9	4.4	2.4	10.6	-10.8	10.9
Mean	-1.4	-0.16	-4.5	-4.5	-0.52	1.3	0.3	1.8	-0.54	0.97
SD	17.4	10.6	8.0	7.7	6.9	8.2	4.4	9.12	9.6	9.13
WX	18		15		23		18		31	

See footnote to table I.

half was processed as a complete response. Due to space limitation, results of 4 subjects are shown in figure 1.

Visual inspection of the results of subject DL7500 suggest (a) no significant difference between the short- and long-run, and (b) no significant difference between the two halves of the long-run. Therefore, other volunteers were asked to perform the long-run only and the two halves were then compared. Results of the other 3 subjects shown in figure 1 also suggest no difference between the two halves.

To quantify the above observations, nonparametric statistics were used because of the relatively small samples. Wilcoxon's signed rank test and the Spearman's coefficient of rank correlation test were used to study population differences and correlating the two halves [2].

Results showed no significant difference between the two halves of the long run ($p > 0.05$) as in tables I–III. This implies that the basic statistics of both halves are similar. Results also demonstrated high correlation between the two halves ($p < 0.05$) except for the phase measure

Table III. Normal population: Gain

Subject	Frequency (1st and 2nd half)									
	0.01 Hz		0.02 Hz		0.04 Hz		0.08 Hz		0.16 Hz	
1	0.3	0.3	0.7	0.6	0.7	0.7	1.0	1.0	0.8	0.8
2	0.2	0.3	0.5	0.4	0.7	0.5	0.8	0.9	0.8	0.8
3	0.2	0.2	0.5	0.5	0.6	0.6	0.6	0.7	0.7	0.6
4	0.4	0.5	0.5	0.4	0.7	0.6	0.5	0.6	0.5	0.6
5	0.5	0.4	0.8	0.8	0.6	0.5	0.5	0.5	0.5	0.4
6	0.3	0.3	0.5	0.6	0.7	0.8	0.7	0.7	0.8	0.6
7	0.3	0.4	0.5	0.4	0.5	0.5	0.6	0.6	0.5	0.6
8	0.3	0.3	0.5	0.3	0.4	0.5	0.5	0.4	0.5	0.4
9	0.5	0.3	0.5	0.4	0.4	0.5	0.5	0.5	0.5	0.6
10	0.4	0.4	0.5	0.5	0.5	0.4	0.6	0.7	0.5	0.6
Mean	0.34	0.34	0.55	0.49	0.58	0.56	0.63	0.66	0.61	0.6
SD	0.11	0.08	0.11	0.14	0.12	0.12	0.16	0.18	0.14	0.13
SCC	0.65		0.75		0.68		0.93		0.76	
WX	22		13		22		15		24	

See footnote to table I.

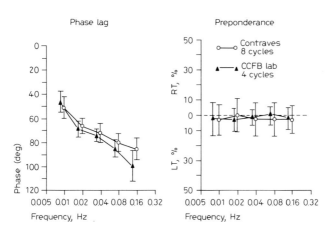

Fig. 2. Results of superimposing short-run and long-run normative data. Each point represents mean ± 1 SD.

at 0.01 Hz. This is because of: (a) using two samples to form the ensemble average and the ensemble standard deviation, and (b) the possible inherent variability of the VOR phase measurement at that frequency.

Figure 2 shows our normative data (50 subjects, short-run test) superimposed on Contraves data (50 subjects, long-run test). Each data point is represented by the mean value ± 1 SD.

The similarity between the two sets of data is evident which further supports the use of the short-run mode with SHA testing. Furthermore, the short-run data display less variability particularly at the mid-frequency range. This is likely because Contraves normative data was collected on different system, although *Wolfe* et al. [1] used the same torque-motor. It is, therefore, important that users should develop their normative data.

The gain measure, although calculated on Contraves system, is not utilized either clinically or to indicate the validity of the vestibulo-ocular reflex response (for example, by calculating the coherence function). In our lab, we use the gain and raw data (so far, qualitatively) to assess response validity. However, the importance of gain and its clinical utilization warrants further studies.

References

1   Wolfe, J.W.; Engelken, E.J.; Kos, C.M.: Low frequency harmonic acceleration as a test of labyrinthine function. Basic methods and illustrative cases. ORL Dig. *86:* 130–142 (1978).
2   Mack, C.: Essentials of statistics for scientists and technologists (Heinemann Educational Books, London 1969).

M.A. Hamid, PhD, DIC., Director, Vestibular Lab.,
Department of Otolaryngology and Communicative Disorders,
Cleveland Clinic Foundation,
9500 Euclid Avenue, Cleveland, OH 44106 (USA)

Adv. Oto-Rhino-Laryng., vol. 30, pp. 171–173 (Karger, Basel 1983)

Harmonic Acceleration Tests for Measuring Vestibular Compensation

Wallace Rubin

Department of Otorhinolaryngology and Biocommunication, Louisiana State University School of Medicine, Metairie, LA, USA

The harmonic acceleration test is not quite the old rotation test [1]. There are some new wrinkles added to the use of the rotation stimulus [2]. These new wrinkles plus the data reduction capability allow us to quantify and measure significant output parameters. These outputs are a result of vestibular stimulation due to body rotation.

Computer technology has and is being applied to the measurement of nystagmus. All of the computer applications are being used for the purpose of data reduction of nystagmic responses from the same old tets (OPK, pendulum tracking, bithermal calorics and conventional rotation tests). Harmonic acceleration tests utilize computer capability for the purpose of adding some new measurable parameters which are practical for patient care. The essence of the harmonic acceleration test is the ability to plot eye movement versus body movement as a result of sinusoidal rotation [3].

The major clinical advantage of harmonic acceleration testing is the monitoring capability [4]. The test results can answer the following questions: (1) Is the patient really getting well? (2) Will the patient get well? (3) Has the patient gotten well? (4) Can we predict which patient will improve? (5) How far along the way is the improvement? (6) Has there been no change at all?

There are other advantages [5]: (1) The stimulus is repeatable, specific and controllable. (2) Drugs and alerting cause very little problem.

(3) A hard copy is immediately available for interpretation, charting and storage. (4) There is no art to reading the recording. (5) Immediate differentiation of normal, peripheral and central pathology is possible. There are also some disadvantages: (1) Right, left differentiation is not always possible. (2) The initial cost is significant. (3) There needs to be a specific commitment of space [6].

Whereas it is always useful to be able to objectively monitor patient progress, it is most important for the clinician to be able to do this in patients who have sustained injury due to trauma. The decisions based on the objective evaluation of the vestibular system are important to the patient, the physician, the employer and many others involved in the situation. Harmonic acceleration testing has made it possible to evaluate progress and make unequivocable decisions regarding occupations and avocations. It has even made it possible to predict progress and in some instances to make intelligent guesses regarding the time frame of return to daily activities and employment. This is indeed a giant step forward.

In the patients with biochemical and metabolic etiologies, confirmation and assessment and evaluation of progress are important in patient management. The correlation of biochemistry and harmonic acceleration findings allows the physician to effectively convince the patient about concurrence with the therapeutic regime and in this way obtain compliance with recommendations. These capabilities for monitor are to the patient's benefit as the physician can be confident about his decisions and therefore more effective in accomplishing therapy goals.

References

1   Mathog, R.H.: Testing of the vestibular system by sinusoidal angular acceleration. Acta oto-lar. *74:* 96–103 (1972).
2   Wolfe, J.W.; Engelken, E.J.; Kos, C.M.: Low frequency harmonic acceleration as a test of labyrinthine function: basic methods and illustrative cases. ORL Dig. *86:* 130–142 (1978).
3   Rubin, W.: Symposium on low frequency harmonic acceleration, the rotary chair, SHA as a modality for monitoring patient progress. Laryngoscope *91:* 1282–1284 (1981).
4   Rubin, W.: The SHA test in vestibular diagnosis. Laryngoscope *91:* 1702–1704 (1981).

5   Wolfe, J.W.; Engelken, E.J.: Olson, J.W.; Kos, C.M.: Vestibular responses to bither-
    mal caloric and harmonic acceleration. Ann. Otol. Rhinol. Lar. *78:* 861–867 (1978).
6   Rubin, W.: Proceedings of the seminar on sinusoidal harmonic acceleration and
    computerized optokinetic tracking tests. Ann. Otol. Rhinol. Lar. *90:* suppl. 86, pp.
    18–24 (1981).

W. Rubin, MD, Clinical Professor, Department of Otorhinolaryngology and
Biocommunication, Louisiana State University School of Medicine,
3333 Kingman Street, Metairie, LA 70002 (USA)

Adv. Oto-Rhino-Laryng., vol. 30, pp. 174–176 (Karger, Basel 1983)

Compensatory Eye Movements during High Angular Accelerations in the Monkey

Andreas Böhmer, Volker Henn

Neurological Department, University Hospital Zürich, Switzerland

Transfer characteristics of the vestibulo-ocular reflex (VOR) are well documented in various species for sinusoidal stimulus frequencies up to 1 Hz. Natural head movements, however, have frequency components up to at least 7 Hz. Using a special primate rotating chair capable of angular accelerations up to 12,000 degrees/s^2 we investigated compensatory eye movements in the horizontal and vertical plane over a frequency range between 0.5 and 6 Hz.

Alert rhesus monkeys, chronically prepared with silver-silver chloride EOG electrodes to monitor eye position, were seated and restrained in a lightweight primate chair with the head firmly fixed by implanted bolts. A device driven by a powerful servo-controlled torque motor rotated sinusoidally either the whole monkey, the head only or the trunk only with the head earth-fixed. Different peak-to-peak amplitudes between 6 and 48 degrees were tested. All experiments were done in light and in complete darkness. Thus the effects of visual, vestibular and neck inputs could be differentiated.

In the *horizontal* plane the VOR nearly perfectly stabilizes gaze up to 6 Hz with a gain slightly below unity and small phase lags increasing with higher frequencies (fig. 1). These findings agree well with data obtained in cats [*Donaghy*, 1980], whereas in the two reports on monkeys [*Furman* et al., 1982; *Keller*, 1978] a gain peak of 1.2 was found at 4 Hz with stimulation in the dark. The *vertical* gain is lower, while the same slight phase lag is found as in the horizontal plane (fig. 1).

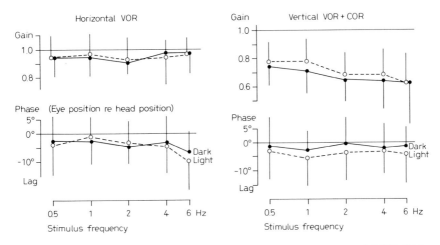

Fig. 1. Bode plots of horizontal and vertical VOR (●) and VOR and OKN (○) in normal monkeys (means and 1 SD of three different stimulus amplitudes, averaged from 5 monkeys).

Experiments were extended to include monkeys with vestibular lesions. The horizontal canals were plugged by drilling through the bony duct and filling the hole with bone chips, an operation kindly performed by Dr. *J. Suzuki*, Tokyo. Rotation was performed in a plane orthogonal to the plane of the intact vertical canals (head pitched 33 degrees forward relative to the stereotaxic plane). The total abolition of inputs from the semicircular canals could be demonstrated by the absence of vestibular nystagmus in response to angular velocity steps of 100 degrees/s. As expected, 7 months after the operation, a cervico-ocular reflex (COR) is established, its gain averaging about 0.3, with a phase lag increasing with higher stimulus frequencies (fig. 2). Sinusoidal rotation of the whole animal elicits eye movements which are phase-advanced and have a gain up to 0.7 at maximum stimulation. In labyrinthectomized monkeys no such responses were seen; therefore somatic afferents can be excluded. We consider the otolith organs to be the most probable candidates for this response. Stimulation of these afferents together with the COR by rotating the animal's head only elicits eye movements in phase with the head movement, the gain being composed of a vectorial summation of the gain of the cervico-ocular and a presumed otolith reflex.

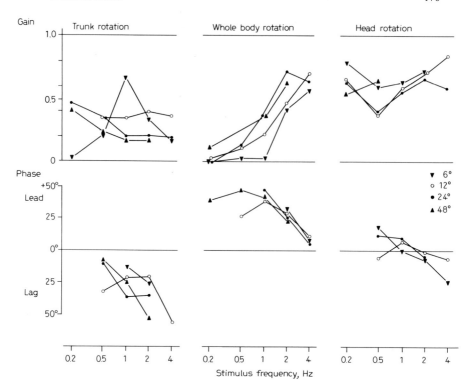

Fig. 2. Bode plots of horizontal VOR and COR of a monkey 7 months after plug-ging both horizontal semicircular canals. Different symbols represent different peak-to-peak stimulus amplitudes.

References

Donaghy, M.: The cat's vestibulo-ocular reflex. J. Physiol., Lond. *300:* 337–351 (1980).
Furman, J.M. et al.: Dynamic range of the frequency response of the horizontal vestibu-lo-ocular reflex of the alert Rhesus monkey. Acta oto-lar. *93:* 81–91 (1982).
Keller, E.L.: Gain of the vestibulo-ocular reflex in the monkey at high rotational fre-quencies. Vision Res. *18:* 311–315 (1978).

Dr. A. Böhmer, Neurological Clinic, University Hospital,
CH–8091 Zürich (Switzerland)

Adv. Oto-Rhino-Laryng., vol. 30, pp. 177–179 (Karger, Basel 1983)

Differential Diagnosis of Acoustic Neurinomas

P. Piffkó, I. Gádor

National Institute of Neurosurgery, Budapest, Hungary

In the present report we have analyzed the last 100 consecutive, unselected pontocerebellar tumor cases of the National Institute of Neurosurgery in Budapest. Breakdown of this total according to histology was as follows: 94 acoustic neuroma, 2 cholesteatoma, 2 meningioma, 1 brainstem tumor and 1 arachnoid cyst.

Let us first examine more closely data from the 94 acoustic neuroma patients. Operations were performed on 29 male and 65 female patients, ranging between 20 and 74 years of age with an average of 48. Lesions were almost equally distributed between the two sides (49 right, 45 left) and no preponderance could be observed. The time elapsed between the onset of first symptoms and between the first admission varied from a low of 2 months to a high of 27 years. On average, 2.2 years passed before our patients felt motivated enough to seek a doctor's advice.

Average CSF protein content was 179 mg/100 ml and only in 6 cases was the protein profile less than 50 mg/100 ml. With respect to size, no small, 30 middle (less than 2.5 cm in diameter) and 64 large tumors were found. 21 cases from the latter group were also accompanied by papilla edema. The frequency of occurrence regarding first symptoms was as follows: 43 progressive sensorineural hearing loss, 30 tinnitus, 13 dizziness, 5 headache, and 3 visual disturbances. 47 of the 94 neuroma patients having total hearing loss at first examination: 38 were presented with severe sensorineural hearing loss; 7 patients had a combined hearing loss, and only 2 had slight hearing deficits.

It has to be emphasized that there was no correlation between the size of the tumor and the degree of hearing loss. Spontaneous nystagmus was absent in 23 patients. Vertical was present in 9, rotatory in 4, directional changing in 10. Horizontal to the affected side in 26, to the unaffected side in 22. In 54 cases the caloric response was absent, in 30 significantly reduced, in 8 cases we found a directional preponderance and in 2 cases the caloric response was symmetrically normal. X-ray examinations have revealed widened internal auditory meatus in 64 patients, while plain X-ray was normal in 30. Of the 34 meatocisternographies only 1 was false-negative. More recently computer tomography (CT) was utilized, and of 27 cases assessed by CT, only 2 small tumors escaped detection. Comparing the data presented above with those of the 6 non-acoustic pontocerebellar tumors the following conclusions can be derived: First symptoms in the 2 meningioma patients were diplopia and a moderate sensorineural hearing loss. Spontaneous nystagmus was not present and caloric response was normal. Both internal auditory meatus were symmetrical on X-ray pictures. In 1 patient cisternography was not performed, while in the other the tumor was shown by contrast examination. In 2 cases of cholesteatoma, 1 patient had headache and dizziness as first symptoms, while the other presented with hemifacial spasms. Hearing was normal in both patients. 1 had no spontaneous nystagmus and the other had gaze nystagmus. Caloric response was symmetrically normal. On Stenvers X-ray the internal auditory meatus was also symmetrically normal. Tumors were verified by meatocisternography in both patients. First symptoms of the patient presenting brainstem tumor were sixth nerve palsy and trigeminal paresthesias, sensorineural hearing loss and rotatory nystagmus to the direction of the unaffected side. Vestibular caloric response was normal. X-rays showed nothing remarkable, meatocisternography did not prove the existence of tumor.

The patient presented with an arachnoidal cyst which was rather difficult to evaluate since a radical mastoidectomy had been performed previously due to a middle ear cholesteatoma. First symptoms were headache with severe conductive hearing loss due to the previous operation. Spontaneous nystagmus was not present and caloric responses were symmetrical. The internal auditory meatus were symmetrical; however, a positive meatocisternography was performed.

The group of 94 acoustic neuroma cases contains 2 patients who were evaluated on the basis of our usual routine examinations as non-acoustic tumors with pontocerebellar localization. Otoneurologic evidence supporting this diagnosis was normal caloric responses and virtually no hearing deficits. At surgery, however, acoustic tumors were found, localized medially with close adherence to the brain stem. These two 'false-negative' cases, with respect to qualitative diagnosis of the tumor, represent a mere 2 % of our material.

P. Piffkó, MD, National Institute of Neurosurgery, Budapest (Hungary)

Adv. Oto-Rhino-Laryng., vol. 30, pp. 180–182 (Karger, Basel 1983)

Significance of Neuro-Otological Examination in Intracranial Diseases

Michihiko Nozue[a], *Kanae Nakamura*[a], *Hiroyuki Mineta*[a], *Ichiro Shimoyama*[b], *Kenichi Uemura*[b]

Departments of [a]Otolaryngology and [b]Neurosurgery, Hamamatsu University School of Medicine, Hamamatsu, Japan

Marked advances have been made in the diagnosis of intracranial diseases since the development of computerized tomography (CT). However, the functional state of the brain and the degree of its damage cannot be evaluated by the conventional CT, and some lesions, particularly in the brainstem, are not detected by CT. Furthermore, classical neurological examination often fails to identify lesions in intracranial diseases. Recent advances in neuro-otology have made it possible to analyze the pathophysiological state of the brain, especially the vestibular and oculomotor systems. Here we would like to stress the significance of neuro-otological examination for intracranial diseases by showing some of our clinical cases.

Case 1: A 54-year-old man complained of tinnitus and hearing loss on the left side. CT demonstrated a cerebellopontine angle tumor on the left. Neuro-otological examinations showed left sensorineural hearing loss, gaze nystagmus and decreased caloric response of the left ear. However, since optokinetic nystagmus (OKN) and eye-tracking test (ETT) were almost normal, we supposed that his brainstem and cerebellar function were preserved normal in spite of the cerebellopontine angle tumor. He was operated and a large cystic acoustic schwannoma of the Anton B type was found and removed.

Case 2: A 47-year-old man complained of right hearing loss, tinnitus and dizziness. In neuro-otological examinations, right sensorineural hearing loss, decreased caloric response, dominant gaze nystagmus, and positional and positioning nystagmus were noticed. OKN was strongly inhibited, and ETT showed saccadic and dysmetric eye move-ment. CT demonstrated a cerebellopontine angle tumor on the right, and it was suggest-ed that his brainstem and cerebellar function might have been considerably disturbed. In operation we found a large solid acoustic schwannoma of the mixed Anton A and B types.

As shown in these cases, neuro-otological studies are quite useful to determine our indication for surgery and in preoperative evaluation of the pathological nature of the cerebellopontine angle tumor.

Case 3: A 35-year-old woman complained of severe headache, vertigo and gait dis-turbance. CT showed a high density area in the cerebellum, and AV malformation was confirmed by angiography. Serial neuro-otological examinations were carried out before and after operation. We noticed that OKN was strongly inhibited before operation, but after operation her OKN showed gradual recovery to almost normal. In ETT, eye move-ment was dysmetric and saccadic before operation, which gradually recovered to normal after operation. These objective findings of recovery course were quite parallel with her other clinical signs and symptoms. It can be stated that neuro-otological examination may also be quite useful to see the dynamic changes, especially the recovery of the cere-bellar and brainstem function.

The following discusses the differential diagnosis between cerebel-lar and brainstem lesions, which is quite difficult in many clinical cases. Because of the intimate interconnection between cerebellum and brainstem, cerebellar lesion could well give considerable influences to the brainstem. However, neuro-otological examinations seem to give us some crucial information about the difference between cerebellar and brainstem lesions. For instance, according to our experiences, in positioning nystagmus test, vertical downward nystagmus was noticed upon head hanging in all cases of cerebellar lesions, while very rare in brainstem lesions. Furthermore, in ETT, eye movements seem to be of-ten dysmetric and saccadic in cerebellar lesions, whereas they are quite smooth in brainstem lesions. In OKN there appears to be some differ-ences between typical cerebellar lesions and brainstem lesions, i.e., in cerebellar lesions, eye movement is ataxic and cannot follow quick pursuit. Therefore, OKN is often inhibited and dysmetric. On the con-trary, in brainstem lesions, eyeballs cannot move as in MLF and PRF lesions. So OKN cannot be induced. Further clinical and experimental experiences are required to elucidate these points.

In summary, we think that neuro-otological examinations are quite useful for the following points: (1) In addition to differential diagnosis of vertigo between the inner ear and intracranial origins, the degree and extent of brainstem and cerebellar dysfunction can be evaluated with critical information for neurosurgical indication. (2) They give much information for follow-up evaluation of the progress and recovery before and after treatment of intracranial diseases. (3) To some extent, differential diagnosis between cerebellar and brainstem lesions could be possible by neuro-otological examination.

M. Nozue, MD, Department of Otolaryngology, Hamamatsu University
School of Medicine, 3600 Handa-cho, Hamamatsu 431–31 (Japan)

Adv. Oto-Rhino-Laryng., vol. 30, pp. 183–186 (Karger, Basel 1983)

Ocular Counterrolling during Constant Velocity Roll in Patients with Brainstem Compression

Shirley G. Diamond, Charles H. Markham

Department of Neurology, UCLA School of Medicine, Los Angeles, Calif., USA

For the last few years, we have been looking at ocular counterrolling (OCR) as a means of examining the function of the otolith organs. First, we determined the general range of responses seen in normal subjects [1–3]. We then examined several patients who had unilateral vestibular nerve sections, usually secondary to excision of acoustic neuromas, and learned how the OCR response differs in persons known to have one nonfunctioning labyrinth [2, 3]. We now review these results briefly and then report on a number of patients with tumors involving the inner ear who were tested prior to any surgical intervention, to see how OCR is disrupted by tumors of various sizes, and how compression of the brainstem may affect this response.

Methods

Our subjects are strapped into a motor-driven chair, and photographs are taken of the upper face while the subject is rolled about the naso-occipital axis at a constant velocity of 3°/s to 90° right ear down, then rolled in the opposite direction to 90° left ear down, and brought back through the upright position (trial I). Without stopping, the procedure is repeated for trial II. Acceleration and deceleration during the first and last 20° of rotation is $0.21°/s^2$, considered by most to be below the threshold of the semicircular canals. We take photographs at each 10° of roll without stopping the rotation.

Measurements of counterrolling are made with a two-projector apparatus using a superimposition technique. This system is accurate to 0.25°. Each eye is measured independently. Details of apparatus and method are described more fully elsewhere [1].

Measurements of OCR are plotted on a preprinted form on which degrees of counterrolling are indicated on the ordinate and tilt of the chair is labeled on the abscissa.

Negative numbers below a zero baseline represent eye torsion to the left, which is the normal counterrolling response to tilt to the right. Positive numbers on the ordinate refer to eye torsion to the right, or what is usually seen when a subject is tilted left ear down.

A solid line printed on the form indicates the mean OCR of both eyes of 16 normal subjects, ranging in age from 9 to 60, as they are tilted through the two trials. When a typical normal subject is tilted to right, his eyes rotate to the left, reaching a maximum amplitude of about 6° of counterrolling when he has been tilted to 60°. Similarly, his eyes roll about 6° to the right when he is tilted left ear down, i.e. OCR is approximately symmetric to both directions of tilt. The plot, or profile, is smooth and trial II looks very much the same as trial I.

A gray band surrounding the mean represents 1.5 SD on either side. When an individual subject's OCR is plotted on this form, the right eye is indicated in green and the left eye in red, and both are superimposed on the normative data, clearly revealing any departure from the usual normal pattern.

Results

First, normal subjects show *consistency* of OCR response in trial II compared to trial I. Second, the two eyes are *conjugate* most of the time. Third, the profile is generally *smooth,* and fourth, counterrolling is approximately *symmetric* with tilt to right or left side.

We have not seen any kind of eye skew or deviation from the primary position in either normal subjects or any of our patients as they were tipped about a naso-occipital axis approximately in the Reid horizontal plane. Following our study of normal subjects, we analyzed data from 8 patients who had undergone unilateral vestibular nerve sections secondary to excision of acoustic neuromas confined to the area of the eighth nerve. These patients showed characteristic profiles of counterrolling, in which *tilt to the intact side* produced abnormal counterrolling while tilt to the lesioned side was relatively normal. The particular kind of abnormality differed between patients, with inconsistency of response to contralateral tilt being the most reliable indicator.

Having thus examined control subjects to establish normative data, and having looked at patients postoperatively whose lesions were clearly known from their surgical reports, we turned our attention to testing patients who had diagnosed tumors, but who had not yet undergone surgical or other intervention. Their OCR profiles were quite easily classified into three groups:

(1) almost normal in appearance; (2) moderately abnormal, and (3) severely abnormal:

(1) Patients with small, purely intracanalicular tumors showed patterns of very mildly disrupted OCR profiles, the abnormality evident when tilted to the side opposite the lesion.

(2) In patients with large acoustic neuromas, 30–50 mm in diameter, involving the internal auditory canal and the cerebellopontine angle, causing some brainstem compression, OCR profiles showed inconsistency in both left ear down and right ear down segments. The profile of OCR when the patients were tilted intact side down was the more abnormal however, generally being inconsistent and lacking in smoothness. We suspect the abnormalities seen in both directions of tilt are the result of brainstem compression in addition to vestibular nerve dysfunction.

(3) Patients with acoustic neuromas which compressed the brainstem to a considerable degree showed profoundly abnormal responses to both directions of tilt. One patient with a right tumor had almost absent OCR when rolled right ear down. When tilted left ear down, her response was opposite the normal one, i.e. her eyes *rolled* rather than counterrolled. Another patient with a large acoustic neuroma which had infiltrated into the left porus acousticus and burrowed a cavity into the left brainstem, showed profound disorganization of OCR with highly aberrant responses to both directions of tilt, his eyes tending to roll rather than counterroll. Both these patients with rolling rather than counterrolling responses prior to surgery showed the same aberrant profiles when examined 4 months after excision of their tumors.

Conclusions

We have tested patients with acoustic tumors ranging from the very small intracanalicular, to the larger tumors impinging on the utricle or utricular nerve, to the more invasive ones which also compress the brainstem. The disruption seen in the counterrolling profiles seem to correspond quite well to the magnitude and location of the tumors. The small intracanalicular tumor affects the counterrolling response to the degree the utricular nerve is involved. As the tumor further involves the utricle or utricular nerve, we see a pattern quite similar to that observed in our unilateral vestibular nerve section patients, i.e. abnormal response when tilted to the side opposite the lesion. In the case of tumors which not only affect the utricle or utricular nerve but also compress the brainstem, we see profound disorganization of response to both directions of tilt. We have examined patients with large cerebellar tumors compressing the brainstem but sparing the utricular nerve and find these persons to have normal counterrolling. From these findings we infer that a combination of utricular and brainstem involvement leads to a major disruption of the pathways subserving the counterrolling response. The rolling observed in the cases of brainstem compression may be due to dysfunction of the vestibular nuclei on the compressed side and of yet unidentified paths, possibly inhibitory, from the opposite vestibular apparatus.

References

1   Diamond, S.G.; Markham, C.H.; Simpson, N.E.; Curthoys, I.S.: Binocular counterrolling in humans during dynamic rotation. Acta oto-lar. *87:* 490–498 (1979).
2   Diamond, S.G.; Markham, C.H.: Binocular counterrolling in humans with unilateral labyrinthectomy and in normal controls. Ann. N.Y. Acad. Sci. *374:* 69–79 (1981).
3   Diamond, S.G.: Markham, C.H.; Furuya, N.: Binocular counterrolling during sustained body tilt in normal humans and in a patient with unilateral vestibular nerve section. Ann. Otol. Rhinol. Lar. *91:* 225–229 (1982)

S.G. Diamond, Department of Neurology, UCLA School of Medicine, Los Angeles, CA 90024 (USA)

Adv. Oto-Rhino-Laryng., vol. 30, pp. 187–192 (Karger, Basel 1983)

Computer Analysis of Electronystagmography Recordings in Routine Equilibrium Examinations

Y. Watanabe, N. Ohashi, H. Kobayashi, S. Takeda, K. Mizukoshi

Toyama Medical and Pharmaceutical University, Faculty of Medicine, Toyama, Japan

Electronystagmography (ENG) has become the routine method for clinical diagnosis of nystagmus responses in equilibrium examinations. So far, ENG recordings have been interpreted solely on the basis of a visual evaluation and manual measurement of the pattern of the waves. In order to facilitate a more objective diagnosis and to obtain more detailed information about nystagmus responses, we have developed a computer program for analyzing the ENG recordings of routine equilibrium examinations.

ENG Recordings and the Computer System

A block diagram of this on-line system is shown in figure 1. In order to examine many patients in a short time, we perform ENG examinations by using two ENG devices at the same time. The time constant of ENG recordings is set to 3 s and a high cut filter, with a cut-off frequency of 25 Hz, removes high frequency noise. The four output signals of each ENG consist of horizontal and vertical eye movement, a marker signal and a current line indicating what kind of examination is performed. These are sent directly through the analog-to-digital converter into a PDP 11/34 computer.

The computer program is designed so that two ENG devices are independently available at the same time. The examiner tells the computer what kind of ENG examination is requested by a simple switching of the ENG panel (fig. 2). When one switch is turned on, then the

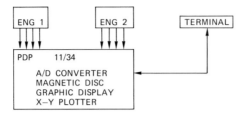

Fig. 1. Block diagram of the computer system.

voltage specified for that particular examination is generated on the line (fig. 3). The start and end points of each ENG examination are signaled by a marker switch and when the marker is set, that specified marker voltage is generated. For communication between the examiner and the computer, a terminal is installed beside the ENG. A cathode-ray tube display and an X–Y plotter are used for the output of the results of analysis.

Operating Procedures

Opening an Examination. An examiner types 'O' from the terminal before the start of an ENG examination and then the computer asks for the patient's name. Only when the patient's name is typed will the computer program for the ENG examination automatically start. After that procedure operation of the terminal is not necessary until the examination is finished.

Calibration. Calibration is performed by having the patients follow the target as it moves sinusoidally, smooth persuit being examined at the same time (eye tracking test, ETT). After starting the ETT the eye movement is recorded on the ENG as sinusoidal waves. An examiner watches the tracing of the ETT and pushes the marker switch when the amplitude of sinusoidal waves is stable. Then the computer points a peak and a valley of a wave and measures the amplitude. The eye movement signal is calibrated by measuring the voltage corresponding to 20 degrees of eye deviation.

ENG Examinations. We routinely perform the following ENG examinations: spontaneous nystagmus, positional nystagmus, caloric

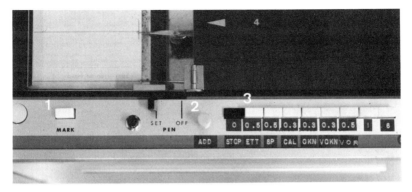

Fig. 2. Switch of the ENG panel. 1 = Marker switch; 2 = special marker switch; 3 = switch indicating what kind of examination is performed (when one switch is turned on the paper speed of the ENG recording is automatically determined).

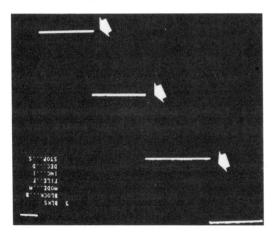

Fig. 3. Voltage generated by switching (indicates what kind of examination is performed).

test, pendular rotation, optokinetic nystagmus. Before the start of one examination, for example, a caloric test, the examiner turns on the switch specifying the caloric test. The paper speed of the ENG is automatically set for the caloric test, i.e. 0.3 cm/s, by that switching. The start and end points of calorization are pointed by the marker switch. In other examinations the procedures are performed as for the caloric test. In case of errors in operation of the marker switch, correction is possible by use of the special marker switch shown in figure 2.

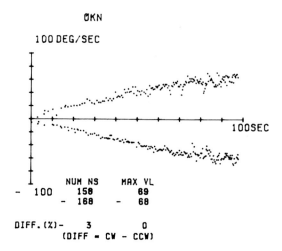

Fig. 4. Results of optokinetic nystagmus.

Data Sampling and Analysis

When an examination starts and the marker signal indicating the starting point of the examination is set, the computer begins data sampling and digitized signals are stored on a magnetic disc. As constant amounts of eye movement data are stored on the magnetic disc, the computer does the analysis of that constant right during the course of the examination. At this stage of analysis only the amplitude and the velocity are computed and stored in the memory.

End of Examination and Output of the Results

When an examination is completed the examiner types 'C' from the terminal. The program computes parameters to evaluate each clinical examination. Computation of the parameters produces the following information: the total number of nystagmus, the maximum slow-phase velocity, canal paresis and directional preponderance percents of caloric test, a gain of vestibulo-ocular reflex (VOR gain) and much more. In clinical use, these analyses are immediately performed after the completion of ENG examinations and the results of the analyses are indicated on the X–Y plotter or the graphic display. The results of

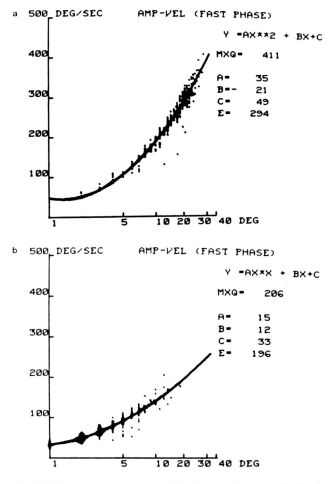

Fig. 5. Relation between the logarithmic translated amplitude of optokinetic nystagmus (abscissa) and fast-phase velocity (coordinate). *a* Normal subject. *b* A case of organic mercury compounds intoxication.

optokinetic nystagmus are shown in figure 4. The graph is a slow phase
velocity series of optokinetic nystagmus printed on the X–Y plotter.
For research purposes, the recorded eye movements are available to
compute more detailed parameters as off-line analysis.

The correlation between the logarithmic translated fast-phase amplitude and the fast-phase velocity in each beat of optokinetic nystagmus can be seen in figure 5a. This is the result of an analysis of a nor-

mal subject. The fast-phase velocity increases indicating parabola as the logarithmic translated fast-phase amplitude increases.

On the graph shown in figure 5b the fast-phase velocity does not increase as steeply as given in figure 5a. This is an analysis of a case of organic mercury compounds intoxication, and the conventional parameters of optokinetic nystagmus indicate a normal value.

Comments

The computer analysis of ENG recordings has been developed and it has been demonstrated that the procedure is useful for the clinical analytic diagnosis of nystagmus responses. Our clinical observations indicate that this method of computer analysis of ENG gives more accurate, objective, detailed and rapid evaluations of nystagmus responses than do conventional manual methods.

References

1   Sills, A.W.; Honrubia, V.; Kumley, W.E.: Algorithm for the multiparameter analysis of nystagmus using a digital computer. Aviat. Space and Environ Med. *46:* 934–942 (1975).
2   Watanabe, Y.; Mizukoshi, K.; Ino, H.: Computer analysis of the electronystagmography. Vestibular mechanisms in health and disease, pp. 131–134 (Academic Press, London, New York, San Francisco 1978).
3   Watanabe, Y.: Computer analysis of ENG recordings in routine equilibrium examinations (in Japanese). Jap. J. Otol. Tokyo 82, suppl., pp. 1553–1578 (1979).
4   Baloh, R.W.; Langhofer, L.; Honrubia, V.; Yee, R.D.: On-line analysis of eye movements using a digital computer. Aviat. Space and Environ. Med. *51:* 563–567 (1980).

Y. Watanabe, MD, Toyama Medical and Pharmaceutical University,
Faculty of Medicine, Toyama 930-01 (Japan)

Adv. Oto-Rhino-Laryng., vol. 30, pp. 193–200 (Karger, Basel 1983)

Electronystagmographic Monitoring in Vestibular Disorders

J.A. McClure

Department of Otolaryngology, University Hospital and University of Western Ontario, London, Ont., Canada

With respect to normal vestibular function, it is well known that the recognition of head motion depends on the generation of asymmetry at a hair cell/first-order neuron level. Asymmetry can also be generated in the peripheral vestibular system by disease. Pathological asymmetry, with respect to semicircular canal input, can be interpreted erroneously by the central nervous system (CNS) as a state of angular motion. On the response side, this asymmetry is manifest by symptomatology such as vertigo and nystagmus. Such nystagmus is called spontaneous nystagmus because it occurs without any form of external stimulation. Peripheral spontaneous nystagmus is easily recognized because it beats continuously in one direction (depending on how the CNS perceives the direction of the asymmetry) and it is accentuated with eyes closed.

In accordance with this concept, a spontaneous nystagmus exists irrespective of the presence or absence of any nystagmus secondary to an induced asymmetry. In fact, spontaneous nystagmus generated by pathological peripheral asymmetry would be additive with respect to other forms of induced nystagmus such as caloric, rotational or positional. Often in the literature, this distinction is not made. For example, *Barber and Stockwell* [2] define spontaneous nystagmus as nystagmus that is direction-fixed and beating with about the same intensity in all head positions when the eyes are closed. However, if the nystagmus changes intensity or direction with position change, then they define it as positional nystagmus. The implication is that spontaneous and positional nystagmus cannot exist simultaneously.

If one accepts peripheral spontaneous nystagmus as an independent entity that signals peripheral asymmetry, then this parameter becomes a tool whereby one can observe the time variant nature of this asymmetry during the course of vestibular disease. In order to measure spontaneous nystagmus one must eliminate all forms of induced nystagmus. This presents no problem except in the case of positional nystagmus where one cannot eliminate the stimulus (i.e. gravity) short of entering a weightless environment. However, based on the assumption that side-to-side gravitational effects are symmetrical with the head in so-called neutral positions (e.g. sitting, supine, caloric test positions), then one can readily observe 'uncontaminated' peripheral spontaneous nystagmus.

A chance observation [7] of spontaneous nystagmus reversal in a patient experiencing an acute attack of Ménière's disease raised the possibility that valuable clinical and pathophysiological information might be contained in patterns of spontaneous nystagmus variation. This lead to the establishment of a system for electronystagmographic (ENG) monitoring of spontaneous nystagmus during acute vestibular upsets. This paper will discuss some of the diagnostic benefits that accrue from this procedure.

ENG Monitor Procedure

In order to collect information on the pattern of spontaneous nystagmus during an acute vertigo attack, it is necessary to initiate ENG monitoring as soon as possible following the onset of symptoms. To accomplish this, patients are given a card with instructions to contact a technician on-call anytime of the day or night should a vertigo attack occur. If it appears that the patient is in the acute stage of an attack, he/she is brought directly to the vestibular clinic in the hospital, and ENG monitoring using DC ENG techniques is commenced immediately. Most patients are admitted to hospital to facilitate their general care since it is often necessary to record the nystagmus intermittently over relatively extended periods. No antivertiginous medications are administered during the monitor period. All recordings are made with eyes closed in a dimly lit room and with the patient in the caloric test position (i.e. head and upper body elevated 30° above the horizontal). Mental alerting is used throughout the recording periods. A calibration procedure is carried out prior to each recording. This requires that the patient gaze back and forth between two red lights that subtend an angle of 20°. Values for the slow-phase velocity of the spontaneous nystagmus are calculated on the basis of the average of 8–10 well-formed beats taken at intervals throughout the trace.

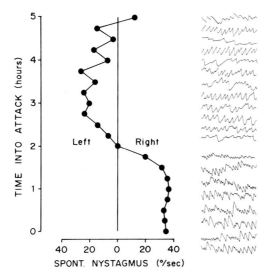

Fig. 1. Pattern of spontaneous nystagmus during an attack of Ménière's disease involving the left side. Initial recording in the acute phase is designated as time zero. Magnitude of the slow phase of nystagmus velocity is plotted on the x-axis. Right and left refer to nystagmus direction (i.e. direction of the fast phase). Actual sample nystagmus recordings for each point (calibrate not the same for each) are shown on the right.

Evidence for Vestibular Involvement

Often with complaints of dizziness, the initial problem is not one of diagnosis but rather a requirement to define the existence of vestibular involvement. The ENG Monitor Program has been useful in this situation since patients can be observed in the clinic when they are complaining of dizziness. Simple physical examination in combination with ENG monitoring readily reveals vestibular involvement in most cases.

Diagnostic Information

The diagnostic information obtained with ENG monitoring is contained in the pattern of progression of spontaneous nystagmus with time. For example, all of our cases of Ménière's disease have shown a pattern of spontaneous nystagmus reversal that reflects the acute and

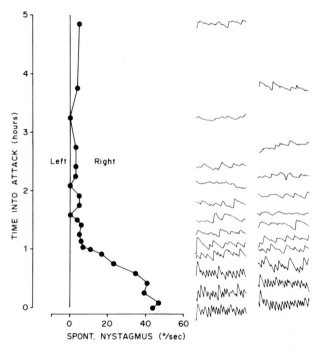

Fig. 2. Pattern of spontaneous nystagmus during a vertigo attack in a patient with recurrent vestibulopathy. Details are the same as those described for figure 1.

recovery stages of the typical attack. An example of such a reversal is shown in figure 1, and the basis for the acute/recovery pattern has been discussed in previous publications [7, 8].

It is interesting, as a further example, to compare the pattern of spontaneous nystagmus in the clinical condition often referred to as vestibular Ménière's disease [1] or more recently designated as recurrent vestibulopathy [6]. This condition is characterized by Ménière-like vertigo attacks without associated hearing loss or tinnitus. Figure 2 illustrates the ENG monitor results from a patient during such an attack. Note the prominent right beating spontaneous nystagmus which beat away from the side with hypoactive function as demonstrated by subsequent caloric tests. As the vertigo subsided, the nystagmus intensity gradually declined toward zero but did not reverse direction. The different pattern of spontaneous nystagmus in this case suggests that the condition is not simply a variant of Ménière's disease.

In terms of other pattern differences that have diagnostic significance, one sees variations in the time scale, repetition of attacks and differences in nystagmus intensity.

Detection of Disease Laterality

ENG monitoring allows one to accurately and objectively define the side with active disease in accordance with the direction of the spontaneous nystagmus. It must be emphasized that an isolated recording of spontaneous nystagmus can be misleading since, as evident in figure 1, this could beat in either direction. However, when the direction pattern with time is coupled with the progression of the patient's vertigo, then the side with active disease becomes obvious.

The value of ENG monitoring in defining the side with active disease is exemplified by 2 of our patients with bilateral deafness since at least early childhood [8]. They both developed incapacitating Ménière-like vertigo attacks during adulthood. By means of ENG monitoring we were able to define the side which was responsible for the attacks, and both were treated successfully by means of a labyrinthectomy in 1 case and an endolymphatic sac decompression in the other.

Most clinicians are aware of cases of so-called failed labyrinthectomy in which the surgery does not completely alleviate the vertigo. Most of these failures are attributed to an incomplete labyrinthectomy, and often the patient will undergo a second operation. In such situations it is possible that the opposite side is the side with active disease and the source of the vertigo. Tests such as the caloric test indicate hypofunction but not necessarily disease activity. ENG monitoring, however, relates function with disease activity, and therefore is able to resolve problems of laterality.

Recognition of Recovery of Function

Jung and Kornhuber [5] and Debain [3] cite the work of Stenger who demonstrated the occurrence of spontaneous nystagmus beating toward the side of vascular and inflammatory lesions, usually within 3 months of the initial problem. Such nystagmus was attributed to recovery of function after central compensation. This has also been our

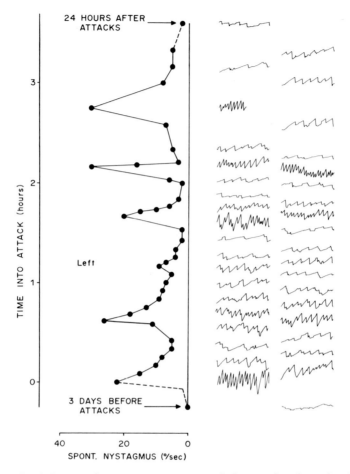

Fig. 3. Pattern of spontaneous nystagmus during a series of transient ischemic attacks. Details are the same as those described for figure 1. The dotted lines indicate that the time of the initial and final points is not to scale.

experience [9] especially following the condition that some call vestibular neuronitis but which we refer to as acute viral labyrinthitis.

Ideally one would always like to be able to show the nystagmus reversal as the patient progresses from acute to recovery stages. In practice, the patient is usually sent for assessment several days or weeks after the acute illness because of persistent low-grade vertigo. The demonstration of a spontaneous nystagmus beating toward a hypofunctioning labyrinth (as determined, for example, by caloric tests) signifies that the patient is probably in a recovery stage. Such information is useful since it allows one to be more objective with respect to prognosis and treatment requirements.

Insight into Pathophysiology

If one accepts that peripheral spontaneous nystagmus reflects the CNSs perception of peripheral asymmetry, then one can gain new pathophysiological insight. The ENG monitor results shown in figure 3 were obtained from a patient during a series of transient ischemic attacks, and these provide an example of the type of pathophysiologic insight one can acquire. By definition, the symptomatology in this condition is caused by loss of function due to neural ischemia, and this is followed by recovery of function when circulation is restored. Despite such recovery, one can see in figure 3 that at no time does the spontaneous nystagmus reverse direction. This absence of recovery nystagmus suggests the possibility that no significant adaptation occurred during the periods of acute asymmetry. Such absence of relatively short-acting adaptation may reflect the more central location of this lesion since in animal experiments [4] such short-acting adaptation has been shown to occur at a peripheral level.

References

1   Alfred, B.R.: Report of Subcommittee on Equilibrium and Its Measurement. Ménière's disease: criteria for diagnosis and evaluation of therapy for reporting. Trans. Am. Acad., Ophthal. Oto-lar. 76: 1462–1464 (1972).
2   Barber, H.O.; Stockwell, C.W.: Manual of electronystagmography; 2nd ed., pp. 145–146 (Mosby, St. Louis 1980).
3   Debain, J.J.: Le nystagmus de régénération. Annls Oto-lar. 91: 691–693 (1974).
4   Goldberg, J.M.; Fernandez, G.: Physiology of peripheral neurons innervating semicircular canals of the squirrel monkey. I. Resting discharge and response to constant acceleration. J. Neurophysiol. 34: 635–660 (1971).
5   Jung, R.; Kornhuber, H.H.: Results of electronystagmography in man: the value of optokinetic, vestibular, and spontaneous nystagmus for neurologic diagnosis and research; in Bender, The oculomotor system, pp. 455–456 (Harper & Row, New York 1964).

6   Leliever, W.C.; Barber, H.O.: Recurrent vestibulopathy. Laryngoscope *91:* 1–6 (1981).
7   McClure, J.A.; Lycett, P.: Recovery nystagmus. J. Otolar. *7:* 141–148 (1978).
8   McClure, J.A.; Copp, J.; Lycett, P.: Recovery nystagmus in Ménière's disease. Laryngoscope *91:* 1727–1737 (1981).
9   McClure, J.A.: ENG monitoring in Ménière's disease and other vestibular disorders. Adv. Oto-Rhino-Laryng. vol. 28, pp. 39–48 (Karger, Basel 1982).

Joseph A. McClure, MD, Department of Otolaryngology, University Hospital, London, Ont. N6A 5A5 (Canada)

Adv. Oto-Rhino-Laryng., vol. 30, pp. 201–204 (Karger, Basel 1983)

Does the Spontaneous Nystagmus Interfere with Eye Tracking?

L. Schalén[a], *I. Pyykkö*[a], *N.G. Henriksson*[a], *V. Jäntti*[b], *C. Wennmo*[a]

[a] Department of Otorhinolaryngology, University Hospital, Lund, Sweden;
[b] Department of Pharmacology, University Hospital, Turku, Finland

Introduction

The results of eye tracking test (ETT) may be affected by spontaneous nystagmus. Thus, slow phases of spontaneous nystagmus may be added to smooth pursuit, resulting in asymmetry of smooth pursuit. Such asymmetry seems to occur more often in patients with spontaneous nystagmus due to disorders of the central nervous system (CNS) than in those with nystagmus due to vestibular end organ lesion [*Baloh* et al., 1977, 1981]. To our knowledge, the possibility of a similar affection of eye tracking by fast phases of spontaneous nystagmus have not been studied systematically. Therefore, in the present study, we conducted the ETT in patients with spontaneous nystagmus of central or peripheral origin. In every case, velocity of smooth pursuit and frequency of superimposed saccades were studied in relation to the direction of the slow and fast phases of spontaneous nystagmus. The aim of the study was to find out whether also an asymmetry in the frequency of superimposed saccades may be disclosed on ETT and if so whether it can be related either to a disorder within the CNS or to vestibular end organ lesion.

Materials and Methods

The group with spontaneous nystagmus of 'central' origin consisted of 6 subjects with disorders involving the cerebellum and brain stem. The mean velocity of spontaneous nystagmus was 6.3°/s, range 4–12°/s. The group with spontaneous nystagmus of 'peripheral' origin consisted of 6 subjects with a sudden unilateral labyrinthine function loss without any accompanying neurological symptoms (=vestibular neuropathy). The mean velocity of spontaneous nystagmus was 5.3°/s, range 2–7°/s.

Eye movements were recorded with the electro-oculographic technique using an ink-jet writer and filter with upper cut-off frequency of 15 Hz and with a paper speed of 2.5 cm/s. The stimulus eliciting tracking eye movements was a spot of light, 3 cm in diameter, travelling on a horizontal screen in predictable ramps with six constant velocities of 10, 20, 30, 40, 50 and 60°/s. Maximum velocity gain of smooth pursuit (eye velocity/target velocity) and frequency of superimposed saccades were measured at each target velocity and in directions of both slow and fast phases of spontaneous nystagmus. The values were compared with normal values in our laboratory [*Schalén* 1980].

Results

All patients with spontaneous nystagmus of 'central' origin had reduced velocity gain of smooth pursuit; among them 4 patients displayed values reduced equally in both directions and two showed asymmetry. The frequency of superimposed saccades was increased in both directions and without relation to either the slow or the fast phase of spontaneous nystagmus.

In all patients with spontaneous nystagmus of 'peripheral' origin, the velocity gain of smooth pursuit was within the normal range in both directions and without asymmetry. The frequency of superimposed saccades tended to increase bilaterally. Significant increase was found only in direction with fast phase of spontaneous nystagmus.

Discussion and Conclusions

Distinct differences were observed in the results of the ETT between patients with spontaneous nystagmus of 'central' and 'peripheral' origin. In patients with 'central' spontaneous nystagmus there was a manifest decrease of velocity gain of smooth pursuit and increase of superimposed saccades, probably independent of spontaneous nystagmus. In addition, asymmetry of smooth pursuit could occur. On the other hand, patients with 'peripheral' spontaneous nystagmus had normal velocity gain of smooth pursuit but there was an asymmetry of superimposed saccades, i.e. the frequency of superimposed saccades was increased in direction with fast phase of spontaneous nystagmus.

It must be stressed that during the ETT, the visual attention is directed towards the moving target and thus, the two parts of the oculomotor system the smooth pursuit and the saccade – must cooperate in order to perform an adequate tracking. Obviously, in cases with spontaneous nystagmus where the input to the oculomotor nuclei is asymmetric, disturbances of eye tracking may be expected. In CNS disorders, as in the present cases with cerebellar and brain stem lesions, the smooth pursuit is often generally affected in both directions. Conceivably, the fact that pathways conveying the smooth pursuit are crossed [*Pyykkö* et al., 1980] may explain a bilateral affection of smooth pursuit even in unilateral disorders of the CNS. Thus, an asymmetry in the velocity gain of smooth pursuit found in some cases – 2 patients in the present study, cases described by *Baloh* et al. [1981] – may be a sign of asymmetric affection of oculomotor structures.

In the patients with vestibular neuropathy, we did not observe any asymmetry of smooth pursuit in relation to slow phase of spontaneous nystagmus, contrary to findings of *Baloh* et al. [1977]. We believe that such asymmetry may occur in some cases with nystagmus of 'peripheral' origin because there is a possibility of summation of visually guided slow eye movements and vestibular nystagmic slow eye movements. In consequence of our method, however, (calculation of maximum not mode or mean velocity gain of smooth pursuit), shortlasting variations of smooth pursuit velocity might have been overlooked. It must be pointed out here that normal velocity gain of smooth pursuit, even in presence of asymmetry of smooth pursuit, indicates a peripheral rather than a central disorder. Moreover, in view of the present results, asymmetry in frequency of superimposed saccades may additionally suggest a peripheral vestibular neuropathy. Namely, in our patients with vestibular neuropathy the most prominent finding was the significant increase of number of superimposed saccades in direction with fast phase of spontaneous nystagmus. A possible explanation may be that the visually and vestibularly induced saccades may be cooperating directionally during a visual task in subjects with otherwise well-preserved oculomotor control mechanisms. Thus, the increase in frequency of superimposed saccades in combination with normal velocity gain of smooth pursuit should be regarded as a compensatory sign of a slight oculomotor imbalance.

References

Baloh, R.W.; Honrubia, V.; Sills, A.: Eye tracking and optokinetic nystagmus. Results of quantitative testing in patients with well-defined nervous system lesions. Ann. Otol. Rhinol. Lar. 86: 108–114 (1977).

Baloh, R.W.; Yee, R.D.; Honrubia, V.: Eye movements in patients with Wallenberg syndrome. Ann. N.Y. Acad. Sci. *374:* 600–613 (1981).

Pyykkö, I.; Hamid, M.; Matsuoka, J. Schalén, L.: Interpretation of nystagmogram, II. Saccades and smooth pursuit eye movements. Pract. Otolar. Kyoto *73:* 2045–2053 (1980).

Schalén, L.: Quantification of tracking eye movements in normal subjects. Acta Oto-lary. *90:* 404–413 (1980).

Lucyna Schalén, MD, Department of Otorhinolaryngology,
University Hospital of Lund, S–221 85 Lund (Sweden)

Adv. Oto-Rhino-Laryng., vol. 30, pp. 205–209 (Karger, Basel 1983)

Visual Suppression Tests in Diagnoses of Diseases of the Central Nervous System

D. Hydén, B. Larsby, L.M. Ödkvist

Department of Otorhinolaryngology, University Hospital, Linköping, Sweden

Introduction

There is a continuous interaction between the vestibulo-ocular reflex (VOR) and the visual system, in order to stabilize gaze during normal locomotion. If the eyes seek to fixate a moving target while the head is in motion the visual system has to suppress the VOR (visual suppression). The suppression ability uses neuronal structures at different levels of the central nervous system (CNS). Clinical studies show that a decreased ability to suppress vestibular eye movements is seen mainly in patients with lesions in the brain stem-cerebellar area [1, 2]. This paper presents data from patients with lesions involving these structures, and in some cases combined with a peripheral vestibular loss. The patients were investigated with standard otoneurological methods and a broad frequency oscillatory test for visual suppression.

Material and Methods

Patients and their diagnoses are listed in table I. The otoneurological investigation included electronystagmogram (ENG) and a smooth pursuit test with a target velocity of 20°/s. Visual suppression was investigated in three different ways, namely: (1) the light was turned on during the caloric response and the patient was asked to fixate a small target; (2) during a sinusoidal oscillation with a Stille chair at a fixed frequency of 0.3 Hz

Table I. Results

Case No.	Age years	Diagnosis	Calorics	Pursuit	VS 0.3 Hz	VOR gain 2.5-3 Hz	VS sinusoidal 0.5-1 Hz	VS sinusoidal 1-1.5 Hz	VS sinusoidal 1.5-2 Hz	VS random 0.5-1 Hz	VS random 1-1.5 Hz	VS random 1.5-2 Hz
1	57	Wallenberg dx	low dx	n	p	0.78	++	++	++	++	n	n
2	58	Wallenberg sin			p	0.90	++	++	+	++	+	+
3	57	Wallenberg sin		p	p	0.88	++	++	+	++	+	+
4	65	Trigeminal neuroma		p	p	1.03	++	++	++	++	++	++
5	13	cerebellar tumor		p		1.27	+	++	++	n	+	+
6	26	cerebellum + BS		p	p	0.50	++	++	++	−	+	+
7	44	cerebellum + BS		p	p	0.61	++	++	++	++	+	n
8	61	Perif + BS	low sin		p	1.28	n	++	++	−	++	+
9	49	Perif + BS	arefl. dx	n	n	0.85	+	++	+	n	n	n
10	69	Perif + BS	arefl. dx	p	p	0.33	++	n	+	n	n	−
11	45	Perif + cerebellum + BS	low dx	p	p	0.56	++	++	++	+	n	+
12	56	Perif + cerebellum + BS	arefl. dx	p	p	0.59	++	++	+	−	n	n

Under 'diagnosis', it is noted whether the disease is a Wallenberg syndrome, or a tumor concerning the cerebellum or brain stem (BS) or if there is a peripheral loss of vestibular function (Perif). p = Pathological; n = normal. For VOR, the gain is given, the normal value being 1.08 ± 0.1. The gain in the broad band tests is given as + or ++ indicating an increase with 1 or 2 SD. A decreased gain is shown by −.

and a peak-to-peak amplitude of 112°, patients were asked to fixate a dot swinging with the chair; (3) with a new broad frequency oscillatory test [3].

The new test is built up around a chair that is driven by a strong hydraulic motor allowing oscillations of the patient over the frequency range 0.5–5 Hz either with a sinusoidal rhythm or following a randomized pattern and thus preventing prediction. Data from monkeys and man with the new test have been published [4, 5]. The broad band test for the present study has been slightly modified [6]. Visual suppression was tested while the patient was asked to fixate a small dot 0.7 m in front of his eyes. The dot was attached to a frame moving with the chair. Suppression was quantified over a frequency area of 0.5–2 Hz. In order to get an indication of the overall VOR function, data were also sampled at 2.5–3 Hz while the patient was fixating a steady target on the wall. The diagnoses based on clinical and X-ray findings were correlated to otoneurological signs (table I).

Results

The findings are summarized in table I. Cases 4, 8, 9 und 11 were investigated before surgery. The total vestibuloocular response at 2.5–3 Hz showed a great variability. In the patients with a peripheral loss it was generally low as expected and in agreement with the caloric response. Two cases with a cerebellar-brain stem lesion (cases 6 and 7) showed a remarkably low VOR gain although their caloric responses were normal. One patient with a pure cerebellar lesion showed a high gain. The Wallenberg cases showed gains slightly below normal, the normal range being 1.08 ± 0.1. Smooth pursuit function were abnormal in most cases. In most cases the visual suppression test during the caloric response was normal. Already with the Stille chair at 0.3 Hz, sinusoidal oscillation in many cases showed a pathological, decreased ability to suppress.

With the visual suppression test using the broad frequency rotatory method with a sinusoidal stimulation all cases w.d.l. showed a pathologically high gain indicating an impaired ability to visually suppress vestibular eye movements.

Random stimulation did not reveal impaired visual suppression to the same extent. This was most evident in the cases with an added vestibular loss. The defective visual suppression was most prominent in the lowest frequency area.

The findings from a patient with a Wallenberg syndrome is displayed in figure 1. It shows a reduced ability to suppress compared to a normal subject.

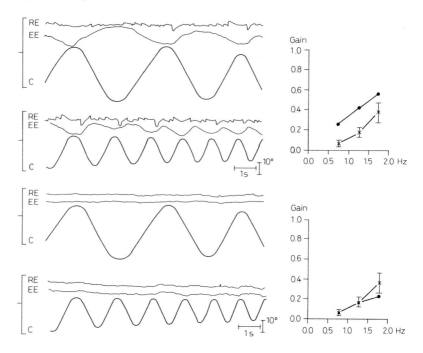

Fig. 1. Recordings from a patient with a Wallenberg syndrome (top) and a normal person (bottom) in the sinusoidal visual suppression test. RE = Raw eye recordings; EE = edited eye recordings, i.e. saccades discarded; C = chair position. To the right gain values are plotted against stimulus frequencies compared to mean values ± 1 SD.

Discussion

It must be emphasized that visual suppression in a rotatory test is highly modified by the ability to predict the stimulus pattern. When prediction is eliminated (random stimulation) the visual suppression ability in normal subjects is markedly reduced compared to during a repetitive stimulation (sinusoidal stimulation). By using both types of stimulation we can test the visual suppression ability both with and without predictive motor programs. As expected, the results from the visual suppression test in the slow chair is almost in total agreement with the findings in the sinusoidal version of the broad frequency test at the lowest frequencies, i.e. 0.5–1 Hz.

In cases with normal peripheral vestibular function visual suppression is also impaired in the random test at low frequencies. If a peripheral vestibular dysfunction is added the visual suppression appears normal while it is still impaired in the sinusoidal test.

Thus, it seems as if in central cases the brain stem/cerebellum area are incapable of generating predictive eye motor programs to suppress the VOR. When predictive motor programs thus are eliminated as in the random test, the lack of visual suppression is not so obvious.

In the light of the literature and the cases reported here it seems reasonable to include a visual suppression test in the routine otoneurological investigation. Our new test allows quantification of visual-vestibular interaction over a wide frequency area during both predictive and nonpredictive stimulation.

References

1   Dichgans, J.; Reutern, G.M. von; Römmelt, U.: Impaired suppression of vestibular nystagmus by fixation in cerebellar and non-cerebellar patients. Arch. Psychiat. NervKrankh. *226:* 183–199 (1978).
2   Halmagyi, G.M.; Gresty, M.A.: Clinical signs of visual-vestibular interaction. J. Neurol. Neurosurg. Psychiat. *42:* 934–939 (1979).
3   Schwarz, D.W.F.; Tomlinson, R.D.: Diagnostic precision in a new rotatory vestibular test. J. Otolar. *8:* 544 (1979).
4   Larsby, B.; Schwarz, D.W.F.; Tomlinson, D.; Istl, Y.: Quantification of the vestibulo-ocular reflex and visual-vestibular interaction for the purpose of clinical diagnosis. Med. biol. Eng. Comp. *20:* 99–107 (1982).
5   Hydén, D.; Istl, Y.; Schwarz, D.W.F.: Human visuo-vestibular interaction as a basis for quantitative clinical diagnostics. Acta oto-lar. (in press, 1982).
6   Larsby, B.; Hydén, D.; Ödkvist, L.M.: A computerized broad spectrum rotatory test. Proc. NES. (accepted for publication, 1982).

D. Hydén, MD, Department of Otorhinolaryngology, University Hospital, Linköping (Sweden)

Adv. Oto-Rhino-Laryng., vol. 30, pp. 210–213 (Karger, Basel 1983)

Abnormal Eye Movement and Nystagmus in the Case of Cerebellar Lesion

Kohji Tokumasu, Naoki Tashiro, Akihiro Ikegami, Satoshi Yoneda, Kiyoko Iho

Kitasato University, School of Medicine, Department of Otorhinolaryngology, Sagamihara City, Japan

Introduction

Abnormal findings of the central type may be frenquently disclosed by neuro-otological examination in the patient with lesions in the cerebellum and/or brainstem. However, the relationship between a location of cerebellar lesion and eye movements still remains obscure, since a lesion in the cerebellum might usually be associated with brainstem lesion, more or less, in clinical cases.

The character of abnormal eye movement due to cerebellar origin was investigated in cases with unilateral cerebellar vascular lesions which had been identified by computed tomography and/or angiography.

Method and Materials

Spontaneous nystagmus, visually induced eye movements and nystagmus induced by vestibular stimulation were examined nystagmographically in cases with cerebellar symptoms, in whom a vascular origin was mostly suspected by physical examination. The test of eye movements was carried out in the acute stage within 1 month after a stroke. Both the location and size of a lesion in the cerebellum were detected by way of computed tomography and vertebroangiography. Finally, 7 cases with cerebellar lesions and with or without the minimum lesions in extracerebellar areas were selected for this study. There were 5 cases with thrombosis, 1 with embolus and 1 with hemorrhage in the cerebellum. In the last case, eye movement was examined after removal of blood clot or cryptic angioma by craniotomy. Occlusion of the superior cerebellar artery was suspected by angiography in 1 case, the anterior inferior cerebellar artery in 3 cases and the posterior inferior cerebellar artery in 2 cases.

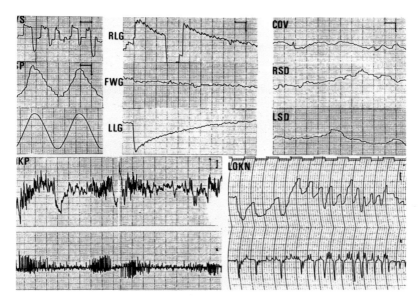

Fig. 1. Pathological eye movement in a case of left cerebellar infarction. A 71-year-old man who had a sudden attack of vomiting, dysarthria, gait disturbance and headache showed a cerebellar sign and deafness on the left side. Dysmetria in voluntary saccades (VS), saccadic eye movement in smooth pursuit (SP), asymmetric gaze directional nystagmus (RLG, FWG, LLG), no nystagmus with eyes covered (COV) and in lateral head position (RSD, LSD) and abnormal optokinetic nystagmus (OKP, LOKN) appeared in the case. Calibration: 10°, 20°/s, 1 s. Time constant: 3 s, 0.03 s.

Result and Discussion

Of 6 males with infarction, who had had hypertension in the past, 1 case (aged 36 years) with cerebellar hermorrhage had not had any particular previous history related to stroke. Sudden attacks of nausea, vomiting, gait disturbance and occipital pain appeared in these cases. (fig. 1, 2). Vertigo or dizziness occurred in 3 cases. Cerebellar signs such as incoordination of the limbs and dysmetria were disclosed in all 7 cases. Other neurological signs were also observed in 5 cases.

The cerebellar sign was dominant in the lesion side in 5 cases, but it was the reverse in 2 cases. Gaze nystagmus and/or spontaneous nystagmus appeared in all 7 cases. 4 cases showed asymmetric horizontal

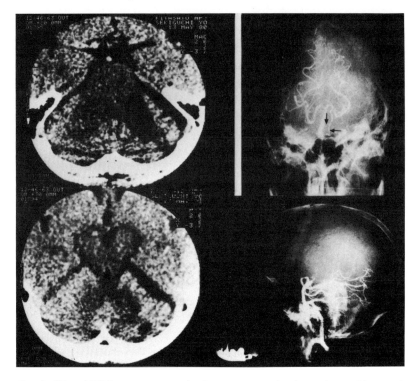

Fig. 2. CT and VAG (same case as in figure 1). Low density and atrophic area is seen in the left half of the cerebellum (CT) and an occlusion of the left anterior inferior cerebellar artery (VAG) is indicated by two arrows.

gaze directional nystagmus. Downbeat or oblique nystagmus appeared in 2 cases. Spontaneous nystagmus towards the intact side and or asymmetric horizontal gaze nystagmus directed dominantly to the intact side were observed in 5 cases. However, just a reversal phenomenon appeared in 2 cases. Overshoot in voluntary saccades, saccadic eye movements in smooth pursuit and elimination of optokinetic nystagmus were observed. Positional nystagmus of direction changing type was discovered in only 1 case. 3 cases showed asymmetric disturbance of both smooth pursuit and voluntary saccade. In such cases, the oculomotor disturbance was dominant when the patient moved his eyes from the intact side to the lesion side. Optokinetic nystagmus towards the lesion side was much more suppressed in 2 cases who showed asymmetric optokinetic nystagmus. Other pathological eye movements such as opsoclonus, ocular myoclonus, flutter-like oscillation and rebound nystagmus, which are considered to be pathological eye movements of cerebellar origin, were not detected.

K. Tokumasu, MD, Kitasato University, School of Medicine,
Department of Otorhinolaryngology, Sagamihara City (Japan)

Adv. Oto-Rhino-Laryng., vol. 30, pp. 214–216 (Karger, Basel 1983)

Neuro-Otological Findings of Spinocerebellar Degeneration with Slow Eye Movement

Atsushi Komatsuzaki

Department of Otorhinolaryngology, Toho University School of Medicine, Tokyo, Japan

Introduction

Spinocerebellar degeneration with slow eye movement is one of the degenerative processes of the central nervous system involving the cerebellum and the brain stem, and the distinctive findings of this disease are disturbances of the eye movement system, especially rapid eye movement such as saccade and quick phase of nystagmus which are more dominant in the horizontal eye movement than in the vertical one. Since this degenerative process has a disposition of heredity, not only the ataxic patient but all family members should be examined both neurologically and neuro-otologically.

The purpose of this report is to evaluate the essential disturbances of eye movement.

Method

Routine neuro-otological examinations including spontaneous and positional nystagmus, pursuit eye movement, optokinetic nystagmus test, caloric test, and auditory brain stem response were performed on four families.

As the distinctive findings of this disease as marked disturbance of saccades or quick phases of nystagmus have already been reported, special interest was placed on the quantitative analysis of peak velocity and duration of these eye movements. The velocity curve was also analyzed with a digital computer set at a sampling time of 3 ms.

Results

Of the four families studied, two had 2 ataxic patients each, and the other two families, 1 ataxic patient each. The general oculomotor findings are summarized as follows: (1) Peak velocity was decreased and duration was prolonged on these eye movements which were induced by visual as well as vestibular stimuli. (2) Optokinetic nystagmus was markedly inhibited on both sides. (3) Instead of producing regular caloric nystagmus, irregular nystagmoid jerks or tonic deviation to the slow phase side was present suggesting disturbance of the rapid eye movement system, whereas the peripheral vestibular system seemed to be intact. (4) Some members of the family showed disorders of saccades or quick phases of nystagmus even though they had no cerebellar signs. After a follow-up study on these members, the general cerebellar signs appeared later. (5) CT scan showed that the atrophy was more dominant on the pons than the cerebellum at a relatively early stage of disease.

Conclusions

Disorders of rapid eye movement are observed in some patients with neurological signs such as Wilson's disease [1], Huntington's chorea [2,3] and spinocerebellar degeneration [4-7]. One distinctive finding of these individuals is that they are unable to produce normal rapid eye movement. Disturbance of rapid eye movement suggests that the lesion must be present in the final common pathway of the horizontal eye movement system. Experimental data show that the essential findings of the lesion of the paramedian pontine reticular formation are disturbance of the rapid eye movement to the ipsilateral side of the lesion [8]. On the other hand, CT scans of some patients demonstrated dominant atrophy was present in the pons. Some patients who had the slow eye movement without the cerebellar signs developed the cerebellar ataxia later. From these data it could be concluded that this degenerative process initially originated from the pontine tegmentum and PPRF plays the important role then spreads to the cerebellum. The neuro-otological approach may contribute to making an early diagnosis.

References

1   Kirkham, T.H.; Kamin, D.F.: Slow saccadic eye movements in Wilson's disease. J. Neurol. Neurosurg. Psychiat. *37:* 191–194 (1974).
2   Starr, A.: A disorder of rapid eye movements in Huntington's chorea. Brain *90:* 545–564 (1967).
3   Avanzini, G.; Girotti, F., et al. Oculomotor disorders in Huntington's chorea. J. Neurol. Neurosurg. Psychiat. *42:* 581–589 (1979).
4   Wadia, N.H.; Swami, R.K.: A new form of heredo-familial spinocerebellar degeneration with slow eye movements. Brain *94:* 359–374 (1971).
5   Starkman, S.; Kaul, S.; Fried, J. et al.: Unusual abnormal eye movements in a family with hereditary spinocerebellar degeneration (Abstract). Neurology *22:* 402 (1972).
6   Singh, B.; Ivamoto, H.; Strobos, R.J.: Slow eye movements in spinocerebellar degeneration. Am. J. Ophthal. *76:* 237–240 (1973).
7   Zee, D.S.; Optican, L.M., et al.: Slow saccades in spinocerebellar degeneration. Archs Neurol. *33:* 243–251 (1976).
8   Cohen, B.; Komatsuzaki, A.; Bender, M.D.: Electrooculographic syndrome in monkeys after pontine reticular formation lesions. Archs Neurol. *18:* 78–92 (1968).

A. Komatsuzaki, MD, Department of Otorhinolaryngology,
Toho University School of Medicine, Tokyo (Japan)

Adv. Oto-Rhino-Laryng., vol. 30, pp. 217–221 (Karger, Basel 1983)

Visual-Vestibular Interaction in Oculomotor Control: A Model Interpretation of Pathological Situations[1]

R. Schmid, A. Buizza,

University of Pavia, Italy

Introduction

Tests combining vestibular and optokinetic stimulations have been used to complete the clinical examination of patients with peripheral vestibular disorders or cerebellar pathologies. The question is whether interaction tests do provide further clinical information with respect to the information that can be obtained from pure vestibular and pure optokinetic tests. In principle, there is the possibility that interaction recruits neural mechanisms that are not activated by a separate stimulation of the vestibular (VOR) and the optokinetic (OKR) reflex. As far as normal subjects are concerned, it has been shown that all types of oculomotor responses that can be evoked in different conditions of VOR-OKR interaction can be closely predicted by models which do not contain any mechanism specific to interaction. The aim of this paper is to show that the same holds for patients with labyrinthine or cerebellar pathologies. The data obtained from each patient during pure vestibular and pure optokinetic tests are used to identify the parameters of a model of VOR-OKR interaction. The model is then used to predict the responses of the same patients in interaction conditions.

[1] Work supported by CNR, Rome, Italy,

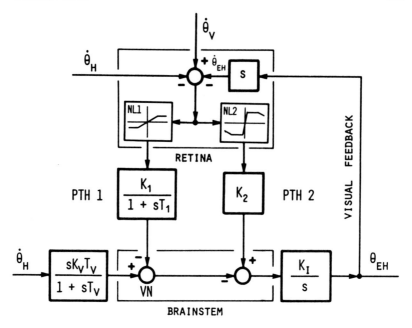

Fig. 1. Model of VOR-OKR interaction.

The Model

The model of VOR-OKR interaction considered in this study (fig.1) has been discussed elsewhere. Its main assumptions are: (a) VOR and OKR interact in a linear way; (b) OKR is composed by two parallel pathways, one (PTH1) passing through the vestibular nuclei (VN), the other (PTH2) passing through the flocculus and reaching the brain stem downstream VN; (c) both PTH1 and PTH2 have nonlinear gain characteristics; (d) the dynamics of PTH1 is that of a low-pass filter with a time constant T_1 of the same order of magnitude as the time constant of the semicircular canals, whereas the dynamics of PTH2 can be neglected. The model in figure 1 has been validated by using experimental data obtained in normals [2]. The nonlinear gain characteristics denoted NL1 and NL2 were identified from optokinetic responses assuming arbitrarily $K_1 = K_2 = K_I = 1$. Since these nonlinearities occur at the level of the peripheral visual system, it is reasonable to assume that they do not change in patients with vestibular or cerebellar pathologies. The variations in the oculomotor responses of these patients to

Table I.

		v30	v60	oc	o30	o60	i30	i60	SI	K_v	K_1	K_2
N	exp	0.42	0.57	0.81	0.75	0.71	0.92	0.88	0.12			
	mod	0.49	0.49	0.81	0.74	0.67	0.88	0.86	0.14	0.5	1.0	1.0
UP	exp	0.49	0.44	0.79	0.84	0.64	0.85	0.83	0.14			
	mod	0.44	0.44	0.76	0.73	0.65	0.86	0.84	0.14	0.45	0.5	1.0
BP	exp	0.07	0.07	0.70	0.68	0.63	0.75	0.77	0.03			
	mod	0.07	0.07	0.71	0.71	0.63	0.73	0.65	0.02	0.07	0.0	1.0

vestibular or optokinetic stimulations can thus be interpreted in terms of changes of the parameters K_v, T_v, and T_1 with respect to their normal values, and of the gains K_1, K_2, and K_1 with respect to unity.

Simulation of Pathological Situations

Two groups of patients were considered, patients with unilateral (UP) or bilateral (BP) horizontal semicircular canal paralysis and patients with different types of cerebellar atrophies. An almost complete set of data on oculomotor responses evoked by pure vestibular, pure optokinetic, and combined visual-vestibular stimulations was available for both groups.

The experimental results for the first group of patients are summarized in table I ('exp') in terms of input-output gains [3]. Eight experimental conditions were considered: sinusoidal oscillation in the dark (v30, v60); constant ($30°/s$, oc) and sinusoidal (o30, o60) optokinetic stimulations; sinusoidal oscillation within an optokinetic drum either stationary (i30, i60) or rotating at a constant velocity ($30°/s$) with respect to the subject (si: superimposition test.) Oscillations of either the subject or the drum had a frequency of 0.05 Hz. Peak velocities were 30 and $60°/s$ except for 'si' tests where peak velocities of 7.5, 15, and $30°/s$ were considered. The results from normals (N) in the same experimental conditions are reported in the first row for comparison. The experimental data obtained from pure vestibular and pure optokinetic tests were used to estimate the model parameters K_v, K_1, and K_2. Since a precise estimation of the time constants T_v and T_1 was not

Table II.

Subject No.	OKN exp	OKN mod	VOR exp	VOR mod	VVOR exp	VVOR mod	K_V	K_1	K_2
1 Friedreich's ataxia	0.65	0.65	0.32	0.32	0.67	0.69	0.33	1.0	0.35
2 Cerebellar degeneration	0.66	0.66	0.61	0.61	0.67	0.83	0.62	1.0	0.37
3 Cerebellar degeneration	0.58	0.57	0.71	0.71	0.90	0.84	0.73	1.0	0.19
4 Spinocerebellar degeneration	0.20	0.20	0.19	0.19	0.29	0.24	0.19	0.46	0.0
5 Spinocerebellar degeneration	0.66	0.66	0.53	0.53	0.69	0.79	0.54	1.0	0.37
6 Olivopontocerebellar degeneration	0.49	0.49	0.52	0.52	0.58	0.68	0.53	1.0	0.08
7 Olivopontocerebellar degeneration	0.49	0.49	0.48	0.48	0.64	0.65	0.49	1.0	0.08
8 Cerebellar degeneration	0.31	0.31	1.02	1.02	0.96	0.93	1.04	0.73	0.0
9 Friedreich's ataxia	0.49	0.49	0.37	0.37	0.63	0.56	0.38	1.0	0.08
10 Cerebellar degeneration	0.41	0.41	0.68	0.68	0.75	0.75	0.69	1.0	0.0
Mean	0.49	0.49	0.54	0.54	0.68	0.70	0.55	0.92	0.15

possible from the available data, normal values were assigned to these parameters ($T_V = 15$ s, $T_1 = 6$ s). The model so defined for each class of patients (UP and BP) was then used to predict oculomotor responses in interaction conditions. The predicted values of input-output gains are reported in the rows of table I denoted by 'mod'.

The experimental results for the group of cerebellar patients are summarized in table II ('exp') [1]. Three experimental conditions were considered: constant optokinetic stimulation at 30°/s (OKN); sinusoidal oscillation in the dark (VOR); sinusoidal oscillation within a stationary optokinetic drum (VVOR). Oscillations were made at 0.05 Hz with peak velocity of 30°/s. After estimation of the parameters K_V, K_1, and K_2 from the results of pure vestibular and pure optokinetic tests, the values of input-output gains reported in the columns 'mod' of table II were predicted by the model.

Discussion and Conclusion

Inspection of tables I and II indicates a good agreement between the values of input-output gains obtained experimentally in interaction conditions and those predicted by the model after estimation of its parameters from the results of pure vestibular and optokinetic tests. Two conclusions can thus be drawn for the groups of patients considered in this paper: (a) VOR and OKR characteristics are the same when the two reflexes are tested independently and during interaction; (b) interaction tests do not provide further information with respect to the information that can be obtained from an appropriate separate testing of VOR and OKR. These conclusions do not exclude some practical reasons for performing interaction tests. First of all, VOR characteristics seem to be more stable in interaction conditions than during rotation in darkness. Secondly, the presence of abnormalities can sometimes be more easily detected by eye inspection in records obtained during interaction tests than during independent VOR and OKR stimulations.

References

1   Baloh, R.W.; Jenkins, H.A.; Honrubia, V.; Yee, R.D.; Lau, C.G.Y.: Visual-vestibular interaction and cerebellar atrophy. Neurol. 29: 116–119 (1979).
2   Schmid, R.; Buizza, A.; Zambarbieri, D.: A non-linear model for visual-vestibular interaction during body rotation in man. Biol. Cybern. 36: 143–151 (1980).
3   Yee, R.D.; Jenkins, H.A.; Baloh, R.W.; Honrubia, V.; Lau, C.G.Y.: Vestibular-optokinetic interaction in normal subjects and in patients with peripheral vestibular dysfunction. J. Otolaryngol. 7: 310–319 (1978).

R. Schmid, MD, University of Pavia, I–27100 Pavia (Italy)

Adv. Oto-Rhino-Laryng., vol. 30, pp. 222–225 (Karger, Basel 1983)

Vestibulo-Visual Interaction Analysis in Normals and Labyrinthectomized Patients

D. Hydén, Y. Istl, D.W.F. Schwarz

Laboratory of Otoneurology, Department of Otolaryngology, University of Toronto, Ont., Canada

Introduction

Diagnostic quantification of vestibular function is usually based on measurements of the vestibulo-ocular reflex (VOR). However, methods available cannot precisely quantify the reflex since either visuo-motor commands, predictive eye movement programmes, proprioceptive input or attenuation to the stimulus can modify the eye movement output. We eliminated these sources of contamination by using a non-attenuation oscillatory rotation stimulus in the plane of the horizontal canal so that proprioceptive input (neck) could not provide systematic errors. Predictive eye movement programmes were ruled out by the use of a randomized oscillation pattern. Contamination due to the eye-tracking systems were eliminated by extending the stimulus dynamics into the high frequency range not covered by those reflexes.

Normal subjects and patients with complete unilateral labyrinthectomies were investigated with a new clinical test for VOR quantification.

Material and Methods

A chair mounted on a strong hydraulic motor (1,356 Nm) permitted horizontal oscillations of a person. Two types of stimulations were used: (a) a pseudo-random oscillation covering a frequency range of 0.5–5 Hz, and (b) pure sine waves at 0.5, 1.0, 2.0, 3.0 and 4.0 Hz. The eye position signal representing the angular position of the stimulus pat-

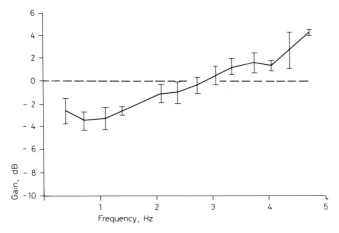

Fig. 1. Gain plot for a fixation suppression task using random stimulation in normal subjects (means ± 1 SD).

tern was fed into a LSI-11 computer for power spectral analysis. The results were presented in Bode plots with the gain of the VOR (eye velocity/head velocity) and the phase angle of the VOR vs. frequency. The validity of the computed values was calculated by a coherence function – for methodology see *Tomlinson* et al. [1] and *Larsby* et al. [2].

Bode plots were obtained for two visual conditions: (1) subjects were instructed to fix a small target on a wall, and (2) subjects were asked to fix a similar target mounted on a frame secured to the chair such that the target moved with the subject (fixation suppression).

Data were collected from 14 healthy adults (age 20–41 years) as well as from 8 patients (age 30–64 years) who had undergone unilateral labyrinthectomy for uncontrollable unilateral Ménière's disease. The time of investigation was from 6 months to 6 years after surgery.

Results

When normals were instructed to fix a stationary target during random stimulation, they produced ideal compensatory eye movements with a gain of 1 (0 dB) between 0.5 and 5 Hz. Patients with unilateral labyrinthine loss had unity gains only at low frequencies, whereas, beyond 2 Hz, there was a gain loss of up to 4 dB. When subjects were instructed to fix a target mounted on the chair during random stimulation, normals showed a gain of −3 dB at the lowest frequencies, which rose to unity at approximately 3 Hz and above unity at higher frequencies (fig. 1).

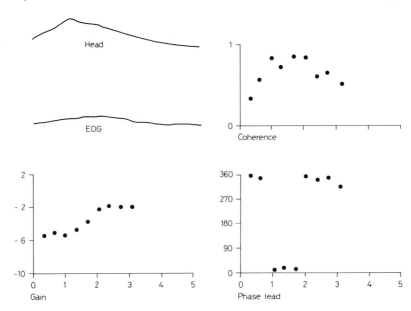

Fig. 2. Graphic display of computation results for one test in which a patient with a right-sided labyrinthectomy suppressed the VOR at random stimulation. Counterclockwise from top left: power spectrum for chair movement and edited EOG respectively, gain plot, phase plot and coherence function.

In the patients, gains were 2 dB below those in normals throughout the frequency range, as examplified in figure 2.

When sine-wave oscillation was used, low frequency gains were much more depressed (to -10 dB); however, gain rise above unity at high frequencies was also observed. In the patients, a significant gain drop was seen only at higher frequencies. In all patients with a unilateral labyrinthine loss, reflex asymmetry could be defined by a linear trend which was superimposed upon the slow compensatory eye movements.

Discussion

We had proposed that it should be possible to isolate the VOR at high frequencies since visual tracking can account for a large proportion of eye position compensation at lower frequencies [2, 3]. However, from our results in normals it is demonstrated that even high frequency gains differ with visual tasks. Thus, predictive eye movement programmes govern fixation suppression as they do smooth pursuit at lower frequencies, and that central programmes can compensate for overestimation of head velocity by vestibular sensors at high frequencies. It therefore seems impossible to isolate the VOR for non-ambigous quantification in alert humans. For practical purposes a fair degree of precision is, however, available at 3 Hz [4]. Confirmation of the correctness of Ewald's second law by the primary canal afferent data of *Goldberg and Fernandez* [5] and by the rotatory nystagmus observations of *Baloh* et al. [6], suggested to us that the compensatory eye movements during head movements should be asymmetrical for patients with unilateral labyrinthine loss. This asymmetry could indicate the side of a peripheral lesion. Our present data confirm that we cannot only quantify the total vestibular loss in a partial case but also detect the side of the lesion.

References

1   Tomlinson, R.D.; Saunders, G.E.; Schwarz, D.W.F.: Analysis of human vestibulo-ocular reflex during active head movements. Acta oto-lar. 90: 184 (1980).

2   Larsby, B.; Schwarz, D.W.F.; Tomlinson, D.; Istl, Y.: Quantification of the vestibulo-ocular reflex and visual-vestibular interaction for the purpose of clinical diagnosis. Med. biol. Eng. Comp. 20: 99 (1982).

3   Barnes, G.R., Benson, A.J.; Prior, A.R.J.: Visual-vestibular interaction in the control of eye movement. Aviation Space environ. Med 49: 557 (1978).

4   Hydén, D.; Istl, Y.E.; Schwarz, D.W.F.: Human visuo-vestibular interaction as a basis for quantitative clinical diagnostics. Acta oto-lar. 94: 53 (1982)

5   Goldberg, J.M.; Fernandez, C.: Physiology of peripheral neurons innervating semicircular canals of the squirrel monkey. I. Resting discharge and response to constant angular accelerations. J. Neurophysiol. 34: 635 (1971).

6   Baloh, R.W.; Honrubia, V.; Konrad, H.R.: Ewald's second law re-evaluated. Acta oto-lar. 83: 475 (1977).

D. Hydén, MD, Laboratory of Otoneurology, Department of Otolaryngology, University of Toronto, Toronto, Ont. (Canada)

Adv. Oto-Rhino-Laryng., vol. 30, pp. 226–229 (Karger, Basel 1983)

Effects of Flocculectomy on Vestibular and Optokinetic Nystagmus and Unit Activity in the Vestibular Nuclei

Walter Waespe, Bernard Cohen, Theodore Raphan

Department of Neurology, Mount Sinai School of Medicine, New York, N.Y., USA; Department of Neurology, University of Zürich, Switzerland, and Department of Computer and Information Science, Brooklyn College, City University of New York, N.Y., USA

Introduction

Optokinetic nystagmus (OKN) is produced by activation of direct and indirect pathways from the visual to the oculomotor system [1]. The direct pathways respond at short latencies to produce rapid changes in eye velocity at the onset and end of OKN. Indirect pathways have the ability to store activity and are responsible for the slower changes in eye velocity during OKN as well as for optokinetic after-nystagmus (OKAN). The indirect pathways are shared in common by the visual and vestibular systems [3]. The location of these pathways is still unknown, but clues as to their nature come from both single unit and lesion studies.

Single cell recordings in the vestibular nuclei of the monkey have shown that vestibular neurons are modulated not only during vestibular nystagmus, but also during optokinetic stimulation [4, 5]. Analysis of this activity indicates that the vestibular nuclei are affected by visual stimuli solely over the indirect pathways [2]. Recordings of unit activity from the flocculus indicate that Purkinje cells carry activity in direct visual oculomotor pathways that mediate rapid changes in OKN slow phase eye velocity [6, 8]. In this communication we present data from lesion and single unit experiments that supports this view.

Methods

Experiments were performed on Rhesus *(Macaca mulatta)* and Cynomolgus *(Macaca fascicularis)* monkeys. Eye movements were recorded with EOG. The EOG was differentiated and rectified to obtain slow phase velocity. Unit activity was recorded with tungsten microelectrodes. The flocculus and parts of the paraflocculus were removed by suction ablation under an operating microscope. During testing monkeys sat in a primate chair under an optokinetic drum. They were given steps of angular velocity about a vertical axis or steps of surround velocity in darkness or in a subject-stationary visual surround.

Results

Vestibular Nystagmus, OKN, and OKAN

The gain and time course of vestibular nystagmus induced by steps of rotation about a vertical axis were unchanged by flocculectomy. This shows that input from the semicircular canals was processed normally over the direct vestibulo-oculomotor and indirect pathways. In contrast, there were distinctive changes in OKN after flocculectomy. Initial rapid increase in slow phase eye velocity at the onset of OKN (fig. 1A, B) was reduced by 60–90% after bilateral flocculectomy (fig. 1C, D). This was found for both vertical and horizontal nystagmus. After unilateral flocculectomy the rapid increase was reduced bilaterally with greater reduction for slow phases of OKN to the side ipsilateral to the lesion. The transition of OKN to OKAN which is normally characterizied by a sudden drop in eye velocity (fig. 1A, B) was smooth after flocculectomy at all stimulus velocities. This indicates that rapid changes in eye velocity induced by movement of the visual surround were lost after flocculectomy. It also indicates that each flocculus contributes to the production of the rapid increase in eye velocity in any direction.

Slow increases in OKN velocity were prolonged after operation, especially at high velocities of stimulation. The longer time course was related to the increased initial retinal slip at the onset of OKN due to the reduction of the rapid rise in OKN. If initial retinal slips were equated, the slow rate of rise was the same before and after flocculectomy. This indicates that the dynamics of the velocity storage mechanism were unchanged by the operation. Steady state levels of OKN were also reduced after operation for stimulus velocities above the saturation level of OKAN (about 50–70° in these animals). However, OKAN velo-

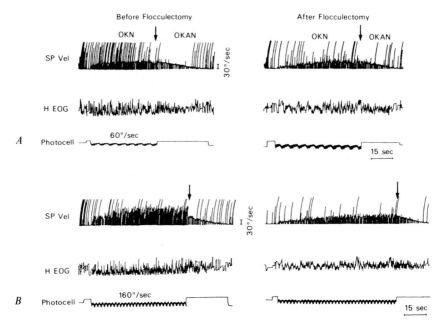

Fig. 1. OKN and OKAN before (left) and after (right) bilateral flocculectomy to velocity steps of 60°/s *(A)* and 160°/s *(B)*. First trace, horizontal eye velocity; second trace, horizontal eye position; third trace, photocell showing duration of stimulation. Downward arrows indicate lights off.

cities and durations were normal showing that the gain of the velocity storage mechanism and the indirect pathways was intact. This is consistent with the view that flocculectomy had affected direct but not indirect visual-oculomotor pathways.

Activity of Vestibular Nuclei Neurons after Flocculectomy

The pattern of activation of vestibular nuclei neurons was qualitatively unchanged by flocculectomy during vestibular stimulation, OKN and OKAN[7]. During OKN vestibular neurons were modulated as before operation; they increased their firing rates proportionally with velocity up to 60°/s, where they saturated as in the normal monkey. The time constant of the increase in firing rate to the steady state level was that of the slow rise of eye velocity. The unit activity declined to its spontaneous level over the time course of OKAN. In the presence of a stationary visual surround, neuronal activity and OKAN were rapidly inhibited and suppressed. The vestibular neurons that were encountered on average had lower gain and sensitivity after than before flocculectomy. However, this appeared to be due to inactivation or loss of neurons that projected their axons to the flocculus rather than to a loss of floccular input to the vestibular nuclei.

Conclusion

The findings support the idea that two mechanisms are involved in the generation of OKN in the primate. Floccular Purkinje cells are part of the direct pathways for the OKN response and do not combine with the indirect pathways at the level of the vestibular nuclei. Vestibular nuclei activity before and after flocculectomy during OKN and OKAN is compatible with the view that the vestibular nuclei are activated by the visual system only over indirect pathways. The finding that flocculectomy does not change the parameters governing the response of the indirect pathways indicates that floccular Purkinje cells do not contribute significantly to their function. In support of this schema the main features of vestibular nystagmus, OKN, OKAN and visual-vestibular interactions were simulated by a model. When the direct visual-oculomotor pathways were removed, the model closely approximated data after flocculectomy [8].

References

1   Cohen, B; Matsuo, V.; Raphan, T.: J. Physiol., Lond. *270:* 321–344 (1977).
2   Raphan, T.; Cohen, B.: in Zuber, Models of oculomotor behavior (CRC Press, 1981).
3   Raphan, T.; Matsuo, V.; Cohen, B.: Exp. Brain Res. *35:* 229–248 (1979).
4   Waespe, W.; Henn, V.: Exp. Brain Res. *27:* 523–538 (1977a).
5   Waespe, W.; Henn, V.: Exp. Brain Res. *30:* 323–330 (1977b).
6   Waespe, W.; Henn, V.: Exp. Brain Res. *43:* 349–360 (1981).
7   Waespe, W.; Cohen, B.: Exp. Brain Res. (in press, 1983).
8   Waespe, W.; Cohen, B.; Raphan, T.: Exp. Brain Res. (in press, 1983).

B. Cohen, MD, Department of Neurology, Mount Sinai School of Medicine, 1 Gustave L. Levy Place, New York, NY 10029 (USA)

Adv. Oto-Rhino-Laryng., vol. 30, pp. 230–234 (Karger, Basel 1983)

Brief Weightlessness and Tactile Cues Influence Visually Induced Roll[1]

Laurence R. Young, Troy A. Crites, Charles M. Oman

Man-Vehicle Laboratory, Massachusetts Institute of Technology, Cambridge, Mass., USA

Introduction

Circularvection, or the perception of self-motion, can be generated about yaw, pitch or roll axes, given the proper rotating visual field stimulus [1–3]. The sensory processing of angular and linear acceleration by the semicircular canals and otoliths, respectively, must be considered in accounting for the phenomenon [4]. The rotating visual field generates involuntary stabilizing eye movements. Rotation of the gravito-inertial force also generates involuntary torsional eye movements [5, 6]. The correlation between the visual stimulus and the perception of motion about the yaw axis, as well as the associated eye movements, has been well documented [2, 3]. In roll vection, however, where the interplay of both the above phenomena are present, the documentation of eye movements and vection properties has been limited. When standing, the observation of a rotating visual field produces a sensation of tilt in the direction opposite to the field rotation [6, 7]. The visually induced tilt is apparently constrained by veridical input from the otolith organs resulting in a paradoxical sensation of continuous motion but constant tilt (roll vection). In the supine position, however, this paradoxical sensation gives way to one of continuous motion about the vertical, since the otoliths no longer provide conflicting information about the gravity vector. The affects of gravity and tactile

[1] This research was supported by NASA Grants NAS 9-15343, NSG 2032, NAG 2-88 and MIT's Sloan Fund for Basic Research.

cues on this self-motion were therefore studied to determine the dependence of this phenomenon upon otolith and semicircular canal inputs, and upon nonvestibular orientation cues. Ocular torsion was also measured to determine its relation to vection.

Methods

A 17-inch diameter dome with a random pattern on its inner surface was rotated about the line of sight of 14 normal subjects under varying conditions. (Nine of the subjects were tested in the zero g airplane, including the 7 science crew members of Spacelab-1, for whom further testing of this nature during extended weightlessness is planned.) Conditions included: 1 g supine, 0 g and 2 g (right ear down). All conditions were tested with and without tactile cues on the subjects' soles. Dome rotation velocities of 30, 45, and 60°/s in two directions (clockwise and counter clockwise) were presented to subjects in random order. For the 0 g tests, a speed of 15°/s was also used. At the beginning and during each trial, eye position was photographed at a rate of 3 frames/s using a 35-mm camera with a circular ring flash, located at the center of rotation. Ocular counterrolling (OCR) photographs were analyzed using iris landmarks and a method described elsewhere [8]. Parabolic flights on NASA's KC-135, providing repeated periods of free fall (25 s duration), were used to evaluate the effects of 0 and 2 g environments on the illusion of vection. The subjects were restrained in a form-fitting mold for protection during flight. The resulting widespread tactile cues may have diminished vection somewhat. Vection onset and velocities were recorded through a subjective position switch indicating 45° of displacement, and were correlated with dome speed, gravito-inertial acceleration, tactile cues and ocular torsion. Tactile cues to the feet were imposed on several runs by having the subjects press against foot restraints in an attempt to create an illusion that the feet were 'down'.

Results and Discussion

The time to onset of vection was shown to be dependent upon all variables tested (fig. 1). The onset latency decreases with increasing stimulus velocity during the 1 g supine test, showing higher velocities to be more compelling stimuli.

In 0 g, the time to onset decreases relative to 1 g supine (significantly at 30 and 45°/s, $p < 0.04$, $t = 2.1$, d.f. $= 13$) and is seen to remain constant over all velocities. When tactile cues are added in the 1 g supine situation they cause significantly longer ($p < 0.01$, $t = 2.6$, d.f. $= 13$) times to onset. They do not, however, represent the equivalent of an upright otolith cue, as this would cause complete elimination of contin-

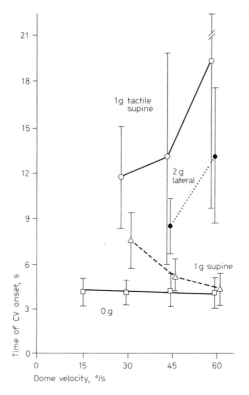

Fig. 1. Time of onset of CV in 4 conditions — 1 g supine, 1 g supine with tactile cues, 2 g lateral (right ear down) and 0 g.

uous vection. A 2 g force directed laterally (right ear downward) also limits onset of continuous vection as compared to 1 g supine. The tactile curve, as well as the 2 g curve, both tend to show an increase in time to onset of CV with increasing dome speed. This is thought to be caused by increasing conflict between visual signals and the nonconfirming otolith system cues. (For the 1 g supine and 0 g situations no such conflict exists.) As the vection stimulus speed increases, more of a conflict is generated between the visual and nonvisual systems and, as a result, onset time for CV lengthens.

Vection velocities extracted from the subjective estimation of successive 45° displacement were shown to be consistently lower than the stimulus velocity. However, these velocities were invariant in all conditions tested.

Torsional eye movements are generated both by rotating visual fields and by changes of the gravito-inertial vector direction [9]. The current measurements, however, showed no further direct correlation to the perceived self-motion. Hering's law of equal innervation was shown to be valid for torsional eye movements; the two eyes rotated through equivalent angles. Dynamic OCR values of 6–10° were seen in most subjects, with no age correlation observed. Torsional nystagmus was noted, but could not be quantified for slow phase velocities due to the low sampling rate.

Conclusions

The decrease in onset time of vection in 1 g supine with increasing visual field speed is in agreement with the findings that increasing stimulus velocities produced shorter onset times in both yaw and roll vection [10, 11]. The decrease in vection onset latency in 0 g, relative to 1 g supine, indicates that the otolith signal is totally ignored during free fall. While otolith information does not present a conflict with vision in a 1 g supine condition, the CNS still receives a valid cue which must be accounted for before self-motion can be accepted. In 0 g, the semi-circular cue is the only vestibular cue denying vection, and it must be overcome before the visually induced self-rotation can be accepted. The latency to onset in 0 g remains constant over all velocities and is approximately 4.25 s. Tactile cues added to 1 g supine presumably create a nonvestibular perception of a local vertical, which conflicts with the visual rotation cues and increases vection latency. Similarly, the 2 g lateral cue conflict further increases the onset latency. The finding that subjective velocities are insensitive to gravity or tactile cues indicates that they are driven by a nonvestibular process similar to an internal clock which locks onto the visual signal and only updates as the scene drastically changes.

The present study indicates that ocular torsion follows the dome rotation and counterrolls to again fix the visual field on the retina. The mean eye position, however, could not be correlated to perceived motion. The induced OCR showed no difference between supine, tactile, nontactile, 2 and 0 g photographs, indicating that the visual drive to OCR is not modified by brief changes in static vestibular stimuli. Binocular recording and viewing showed greater (but not double) amounts of OCR than monocular viewing, again indicating that dynamic OCR is driven by the visual and not the static vestibular input [1]. Although this 1 g supine study did not show a close linkage between ocular torsion and vection, other studies of roll vection in 1 g upright do show a reduction in torsion during vection [12, 13]. Further refinement of a high resolution, high sampling rate (60 Hz or greater) method of OCR measurement is needed to further detail the eye movements seen in visually induced roll [14].

References

1   Troy, A. Crites: Circularvection and ocular counterrolling in visually induced roll – supine and in weightlessness; SM thesis, MIT, Cambridge (1980).
2   Dichgans, J.; Brandt, Th.: in The Handbook of sensory physiology – perception, vol. VIII, chapt. 25 (Springer, Berlin 1978).
3   Young, L.R.: Visual and vestibular influence in human self-motion perception; in Gualtierotti, Vestibular function and morphology (Springer, Berlin 1981).
4   Young, L.R.; Dichgans, J.; Murphy, R.; Brandt. T.: Interaction of optokinetic and vestibular stimuli in motion perception. Acta oto-lar. 76: 24–31 (1973).
5   Miller, E.F.: Counterrolling of the human eyes produced by head tilt with respect to gravity. Acta oto-lar. 54: 479 (1962).
6   Miller, E.F.; Graybiel, A.: Effect of gravito-inertial force on ocular counterrolling. J. appl. Physiol. 31: 697–700 (1971).
7   Dichgans, J.; Held, R.; Young, L.R.; Brandt, T.: Moving visual scenes influence the apparent direction of gravity. Science 178: 1217–1219 (1972).
8   Lichtenberg, B.K.: Ocular counterrolling induced in humans by horizontal accelerations; ScD thesis, MIT, Cambridge (1979).
9   Wolfe, J.M.; Held, R.: Eye torsion and visual tilt are mediated by different binocular processes. Vision Res. 19: 917 (1979).
10 Young, L.R.; Oman, C.M.: Influence of head position and field on visually induced motion effects in three axes of rotation. Proc. 10th Annu. Conf. on Manual Control, pp. 319–440 (1974).
11 Young, L.R.; Oman, C.M.; Dichgans, J.: Influence of head orientation on visually induced pitch and roll sensation. Aviat. Space envir. Med. 46: 264–268 (1975).
12 Finke, R.; Held, R.: State reversals of optically induced tilt and torsional eye movements. Percept. Psychophys. 23: 337 (1978).
13 Young, L.R.; McQuain, M.T.: Relationship between ocular torsion and roll vection in an upright 1 g field (in preparation, 1983).
14 Edelman, E.R.: Video based monitoring of torsional eye movements; SM thesis, MIT, Cambridge (1979).

L.R. Young, ScD, Man-Vehicle Laboratory, Massachusetts Institute of Technology, Cambridge, MA 02139 (USA)

Adv. Oto-Rhino-Laryng., vol. 30, pp. 235–237 (Karger, Basel 1983)

Recovery of Decreased Compensatory Eye Movements and Gaze Disturbances in Patients with Unilateral Loss of Labyrinthine Function

Masahiro Takahashi[a], *Takuya Uemura*[b], *Takehisa Fujishiro*[a]

[a] Department of Otolaryngology, Tokyo Women's Medical College, Tokyo, Japan;
[b] Department of Otolaryngology, Kyushu University School of Medicine, Fukuoka, Japan

In the early stage of unilateral loss of labyrinthine function, patients present a decrease in compensatory eye movement and disturbance of spatial gaze fixation during head rotations to the affected side [1]. In this paper, we show the relationship between recovery of the gaze disturbance and the change in compensatory eye movement in the dark after the onset of the disease.

Subjects and Methods

We report on 9 patients with unilateral loss of labyrinthine function; they manifested no response upon ice water irrigation of the ear on the affected side. There were 5 patients with vestibular neuronitis, 2 with labyrinthitis, and 1 patient each with temporal bone fracture and Ramsay-Hunt's syndrome.

The patients were asked to turn their heads smoothly from right to left in time to a metronome at an amplitude of 40° at three different speeds, i.e. at the rates of 1, 2 and 3 cycles during a 3-second period. Rotation tests were under the two different visual conditions: with the patient performing mental arithmetic in the dark and with the patient visually fixating on a target on the wall. Eye movements were recorded on a DC electrooculograph through bitemporal leads with the eyes open. Horizontal head movements were recorded on a head rotation detector. Since the duration of the observation periods and the times of examination varied among the patients, the mean values of the gain of

2-5 patients were obtained at five different points: at 4 and 10 days, and at 1, 2 and 6 months after the onset of the disease.

Results

Mental Arithmetic in the Dark

Even during the acute stage of the disease (the 4th day after the onset), nystagmus was induced in both directions of head rotation, however, slow-phase eye speed was lower during rotations to the affected than the intact side. Therefore, the gaze recordings became distorted during rotations to the affected side. With respect to compensatory eye movement, the difference between rotations to the intact and the affected side decreased as the interval between the onset and the examination increased. On the 4th day, the mean gain upon ratations to the intact side varied from 0.68 to 0.89 depending on differences in the turning frequency; these values were 0.68–0.73 at 6 months. Although the gain slightly decreased with passage of time after onset, it remained in the range obtained in 20 normal subjects [2]. In the acute stage (day 4) the gain was markedly small (0.13–0.17) during rotations to the affected side. It gradually increased with passage of time and at 6 months, it was 0.52–0.58. These values represent 70–85% of those obtained upon rotations to the intact side.

Visual Fixation on a Lamp on the Wall

In the acute stages, due to an apparent increase in compensatory eye movement, gaze fixation on a target on the wall was slightly impaired even during rotations to the intact side. However, this impairment disappeared almost completely within a month of the onset. On the other hand, during rotations to the affected side, spatial gaze fixation was remarkably disturbed for a long time. Gaze disturbance became more marked as the frequency of head rotation increased; it recovered with time, as did compensatory eye movement in the dark. The mean gain value obtained during rotations to the intact side was 1.14–1.19 on the 4th day; it decreased to 0.99–1.04 at 1 month. Upon rotations to the affected side, the gain showed smaller values as the frequency of head rotation increased; on the 4th day, it was 0.56 at 0.33 Hz and 0.32 at 1.0 Hz. Although at 2 months, the gain showed a value close to unity at 0.33 Hz, it was 0.83 at 1.0 Hz even 6 months after the onset.

Comment

The disturbance of gaze fixation upon rotations to the affected side resembled that found in patients with bilateral lesions, therefore, compensatory eye movement for spatial gaze fixation seems to decrease only during rotations to the affected side. During the time course of compensation, gaze disturbance recovered in parallel with the decrease in the directional difference of compensatory eye movement in the dark; it consisted of a marked increase in gain upon rotations to the affected side and a slight decrease upon rotations to the intact side. During rotations to the affected side, the gain in the dark (0.52–0.58) examined 6 months after the onset was very close to the value obtained in patients with bilateral lesions who were examined 9–15 months after the onset (0.50–0.61). Thus, the relationship between gaze disturbance and the decrease in compensatory eye movement in patients with unilateral lesions may be very similar to that in patients with bilateral lesions, not only in the early, but also the compensated stage of the disease. Recovery of gaze disturbance in unilateral lesions is thought to be accomplished by three mechanisms: initiation of compensatory eye movement during rotations to the affected side, a decrease in the right-left difference of compensatory eye movement in the dark, and recalibration of amplification of compensatory eye movement in the light to obtain spatial gaze fixation.

References

1   Takahashi, M.; Uemura, T.; Fujishiro, T.: Compensatory eye movement and gaze fixation during active head rotation in patients with labyrinthine disorders. Ann. Otol. Rhinol. Lar. 90: 241–245 (1981).
2   Takahashi, M.; Uemura, T.; Fujishiro, T.: Studies of the vestibulo-ocular reflex and visual-vestibular interactions during active head movements. Acta oto-lar. 90: 115–124 (1980).

M. Takahashi, MD, Department of Otolaryngology, Tokyo Women's Medical College, Tokyo 162 (Japan)

Adv. Oto-Rhino-Laryng., vol. 30, pp. 238–241 (Karger, Basel 1983)

Modification of Nystagmus Suppression by Peripheral Location and Strobe Rate of Head-Fixed Targets

G.R. Barnes

RAF Institute of Aviation Medicine, Farnborough, Hants, England

At present, there is little quantitative evidence about the degree to which the characteristics of a visual display modify supression of vestibular nystagmus. In the following experiment, an assessment has been made of the effects of degrading visual information, first by the presentation of the visual stimulus in the peripheral visual field and second by tachistoscopic presentation of foveal and peripheral visual targets [*Barnes*, 1982].

The subject was seated on a large turntable to which his head and body were firmly secured. He viewed a display, also rigidly coupled to the turntable, which consisted of nine red target lights, placed on a periphery in the horizontal plane. Eye movements were recorded using an infra-red recording apparatus (resolution 10–20 min arc) incorporated into a helmet-like system which was rigidly coupled to the head by a bite-bar.

Experiment I

The subject was exposed to sinusoidal angular motion about the yaw axis of the body, at frequencies between 0.25 and 2.0 Hz; peak velocity was maintained at ±60°/s. 6 subjects were each exposed to the following experimental conditions: (a) the subject was instructed to maintain constant fixation on a single central target light; (b) eye movements were recorded in complete darkness; (c) in four separate conditions the subject was presented with a pair of target lights at one of four peripheral locations (±2.5°, ±5°, ±10° or ±20°). In condition c there was no central fixation light but the subject was instructed to look fixedly at the black featureless space midway between the two peripheral target lights.

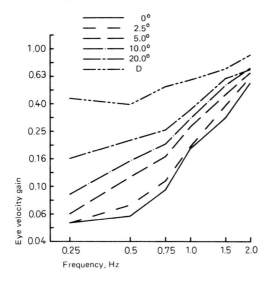

Fig. 1. The ratio of slow phase eye velocity to head angular velocity (gain), as a function of stimulus frequency. 0–20° = target locations; D=darkness.

The most important finding of this experiment was a highly significant (p<0.001) increase in the ratio of slow-phase eye velocity to turntable velocity as the target sources were moved further into the periphery (fig. 1). The effect of the peripheral location of the target changed with the frequency of stimulation. At the lowest frequency (0.25 Hz) even the target lights at ±20° were able to exert a considerable degree of suppression (65% on average), whereas at the higher frequencies only the centrally located targets (0 and ±2.5°) had any significant effect. There was no evidence of any complete cancellation of the vestibular response when viewing the central target. Rather, it appeared that the presence of a fixation point led to a reduction in the amplitude of saccades and an increase of saccadic frequency.

Experiment II

In the second experiment the target lights were illuminated in a tachistoscopic manner. The duration of the light pulse was maintained at 100 µs whilst the inter-pulse interval was varied between 10 and 3,000 ms. The subject was presented with either a centre target light or a pair of lights at ±10° from centre as in experiment I. The stimulus was a sinusoidal oscillation in yaw at a frequency of 0.5 Hz, with a peak velocity of ±60°/s.

In all subjects the degree of suppression of the vestibulo-ocular reflex decreased in a graded manner as the inter-pulse interval was increased (fig. 2). There was a highly significant (p<0.001) increase in the ratio of slow-phase eye velocity to head velocity

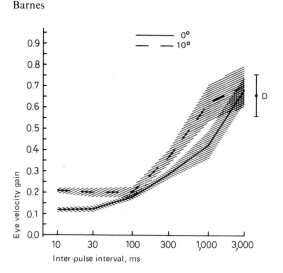

Fig. 2. The effects of inter-pulse interval on the ratio of eye velocity to head veloci-ty during tachistoscopic presentation of head-fixed targets.

with increase of inter-pulse interval. The eye velocity gains obtained at an inter-pulse in-terval of 3,000 ms for both the centre and ±10° target presentations were not significant-ly different from each other, nor were they significantly different from the response in darkness, for which the mean level of eye velocity gain was 0.65.

Discussion

It is apparent from the experimental results that visual feedback of retinal error information is essential in order to achieve optimum sup-pression. Graded levels of suppression are achieved if the retinal error information is degraded by moving the target sources further into the periphery or if the target lights are presented tachistoscopically. There was never any complete suppression of the vestibular nystagmus al-though, during optimal suppression conditions, the slow-phase eye ve-locity was sufficiently low for there to be no significant blurring of the image. In the presence of such a prevailing nystagmus there appears to be a powerful alternative mechanism for foveating the image, which operates by controlling the size and frequency of the saccadic eye movements.

Reference

Barnes, G.R.: The effects of retinal location and strobe rate of head-fixed visual targets on suppression of vestibular nystagmus; in Roucoux and Grommelinck, Physiological and pathological aspects of eye movements, pp. 281–300 (Junk, The Hague 1982).

G.R. Barnes, PhD, RAF Institute of Aviation Medicine, Farnborough, Hants (England)

Experimental and Clinical Aspects of Ménière's Disease and Other Vestibular Disorders

Adv. Oto-Rhino-Laryng., vol. 30, pp. 242–244 (Karger, Basel 1983)

Experimental and Clinical Studies on Epidural Drainage Surgery

Masaaki Kitahara, Taizo Takeda, Hideharu Matsubara, Yoshiro Yazawa, Hitoshi Kitano

Department of Otolaryngology, Shiga University of Medical Science, Otsu, Japan

The drainage of endolymphatic fluid at the endolymphatic sac is one of the widely accepted surgical techniques for the treatment of Ménière's disease. This drainage surgery seems to be the most logical one; however, its basic research is still lacking since the surgery was first performed by *G. Portmann* in 1926. In this study, the effects of opening the endolymphatic sac will be examined by using two types of hydrops models (underabsorption and overproduction) of Ménière's disease. The principle and results of our epidural drainage surgery will also be discussed.

Experimental Studies

(1) The endolymphatic, perilymphatic and cerebrospinal fluid pressure of guinea pigs with hydrops caused by the cauterization of the sac were measured with a Statham P23ID pressure transducer. The endolymphatic pressure was not higher than that of normal guinea pigs. In humans, this tendency is expected to be more manifest since the human cochlear aqueduct has a barrier which is abscent in guinea pigs. We thought a more adequate drainage effect could be obtained with epidural drainage than with subarachnoid drainage since cerebrospinal pressure is not always lower than endolymphatic pressure,

but the atmospheric pressure of the epidural space is always lower than endolymphatic pressure.

(2) When the endolymphatic sac was opened epidurally after hydrops had been made, the Reissner's membrane was so folded back upon itself that the decrease of the hydrops was confirmed. Although the scala media was opened into epidural atmospheric space with 0 mm of water pressure through the vestibular aqueduct, its pressure did not decrease to 0 mm of water. There seems to be some kind of regulating mechanism in the vestibular aqueduct. The sensory epithelium in the organ of Corti appeared to be normal. Also, in electrocochleograms of normal guinea pigs with open endolymphatic sacs, no abnormality was observed.

(3) When methylene blue was applied to scala media with 100 mm of water pressure, it easily passed through the vestibular aqueduct into the sac. When methylene blue was applied to the endolymphatic sac with 300 mm of water pressure, no methylene blue was visible at a small hole made on scala media. The passage through the vestibular aqueduct seems to be one way, i.e., from the scala media to the endolymphatic sac. This finding suggests that inflammation, suppuration etc. around the sac has little influence on the cochlear duct through the vestibular aqueduct.

These findings indicate that epidural drainage surgery exclude any possibility of damage to the cochlear duct.

(4) When artificial endolymph was applied into scala media through the stria vascularis with a speed of 0.02 μl/s, the endolymphatic pressure increased more slowly in the guinea pigs with open endolymphatic sacs than in normal guinea pigs. This effect could be obtained only when the endolymphatic sac opening included the rugous portion. When artificial endolymph was applied with a speed of 0.08 μl/s, however, the effect was not observed.

These findings indicate that when the rugous portion is opened, epidural drainage surgery would be effective unless a vast volume of endolymph is produced.

Clinical Study

Our epidural drainage technique consists of the following 3 principles: (1) the epidural opening of the endolymphatic sac including the rugous portion; (2) the folding back of the lateral wall of the sac, and an insertion of a gel film into it, and (3) undergoing other treatment in combination with the surgery. Our surgical technique is as follows:

After simple mastoidectomy, the endolymphatic sac is exposed including the rugous portion. The rugous portion can be identified by touching the posterior bony wall through it. With a fine knife, a backward L-shaped incision is made as wide as possible in the lateral wall of the sac including the rugous portion. Next, the incised flap is folded back in the direction of the apex, and the corner of the flap is inserted between the dura and the petrous bone in order not to close off the drainage pathway. A long narrow triangualar piece of gel film, which is absorbed in 2 or 3 months, is placed into the aqueduct and sac with the wider end extending out of the sac onto the dura of the posterior cranial fossa. The endolymphatic sac and gel film is then covered with gelatin sponges to promote the formation of a thin membrane and to form pneumatic cells. After this, the soft tissues are closed in the usual way.

The long-term results of 55 patients upon whom our epidural drainage surgery had been performed between 1970 and 1975, were compared with those of 105 patients who had had medical treatment during the same period. After the treatment, these patients were observed for over 5 years. By using our drainage technique, vertigo was completely controlled in 82% and hearing loss of 15 dB or more remained in only 13%. With medical treatment vertigo was completely controlled in only 16% and hearing worsened in 36%. When the long-term results of two series of treatment are compared, it is quite obvious that the results of the special epidural drainage technique are more favorable regarding the relief of vertigo and the preserving of hearing ability.

Based on these experimental and clinical studies, we would like to conclude that our new epidural drainage surgery presently apears to be the most effective and safe surgical procedure for the elimination of vertigo, while preserving hearing ability.

M. Kitahara, MD, Department of Otolaryngology, Shiga University of Medical Science, Otsu 520-21 (Japan)

Adv. Oto-Rhino-Laryng., vol. 30, pp. 245–253 (Karger, Basel 1983)

Relation between Blood Pressure and Plasma Norepinephrine in Diagnosis of Ménière's Disease

Toru Matsunaga, Jun-Ichi Yoshida, Etsuko Tagami, Masuaki Okada

Osaka University Medical School, Osaka, Japan

Introduction

Numerous proposals have been offered for the etiology of Ménière's disease, the pathohistology of which is characterized as the endolymphatic hydrops by temporal bone study [8]. The involvement of sympathetic vasomotor disturbances in inner ear circulation has been accepted to be very important [6, 9]. Recently developed reliable assays of circulating catecholamine levels make it possible to evaluate the sympathetic tone and reactivity in humans [1]. We have investigated plasma catecholamine levels in Ménière's disease patients by using gas chromatography-mass spectrometry and studied the clinical significance of the relation between blood pressure and plasma norepinephrine [5]. These preliminary findings suggested that Ménière's disease patients during orthostasis show the hyporesponsiveness of blood pressure to endogenously released norepinephrine which may be due to a transient and functional disturbance of adrenergic receptor. The present study aimed to verify whether the hyporesponsiveness to endogenously released norepinephrine is solely shown in Ménière's disease or not, and secondary, to examine whether such characteristic pressor response is also valid to exogenously administered norepinephrine or not.

Table I. Subjects studied

	Group A				Group B			
	n	sex M:F	age	blood pressure	n	sex M:F	age	blood pressure
Ménière's disease	23	14:9	42 (23–61)	81.6 ± 12.2	10	4:6	43 (26–56)	88.7 ± 3.9
Sudden deafness	12	8:4	36 (19–49)	86.9 ± 12.4	5	1:4	41 (28–49)	80.6 ± 2.2
Control	10	5:5	28 (19–43)	83.8 ± 10.0	6	4:2	29 (25–36)	83.5 ± 3.1

Subjects and Method

Subjects were divided into groups A and B (table I). Group A consisted of 10 healthy adults (control) and 35 patients, i.e., 23 with Ménière's disease and 12 with sudden deafness. They were tested for the effect of standing upright from supine position on blood pressure and plasma norepinephrine concentration.

Group B consisted of 6 healthy adults and 15 patients, i.e., 10 with Ménière's disease and 5 with sudden deafness. They were tested for the dose-related effect of norepinephrine infusion on blood pressure and plasma norepinephrine concentration.

All the patients with Ménière's disease were off the attack stage and complained of vertigo, tinnitus and hearing loss, showing fluctuating hearing acuity in audiometry. None of the subjects suffered from hypertension, nor did they differ significantly in mean age or mean arterial pressure. They received no medication for 1 week preceding the test.

2 h after lunch the subjects were instructed to assume a supine position after which an indwelling catheter was introduced into an antecubital vein of the dominant side. After 20 min in this position, their blood pressure was measured with a sphygmomanometer and, without delay, the blood was withdrawn in a heparinized syringe. In group A, the subjects then stood upright and had their blood pressure measure and blood collected twice after standing upright for 5 and 10 min. In group B, another indwelling catheter was inserted into the antecubital vein of the nondominant side and was maintained patent by infusion of 5% dextrose in water. The opposite arm was used to record blood pressure at 1-min intervals and to obtain blood samples.

After a baseline blood sample was obtained and the blood pressure measured, L-norepinephrine bitartrate (5 μg base/ml in 5% dextrose in water) was infused initially at a rate of 2.5 μg/min. The infusion rate was increased every 10 min to 5, 10, 15 and finally to 20 μg/min according to *Polinsky* et al. [7]. Blood was taken at the end of each 10-min period. The infusion rate was increased until: (1) the systolic blood pressure increased by 40 mm Hg; (2) the diastolic blood pressure increased by 30 mm Hg, and (3) the pulse decreased by 30 bpm.

All the collected blood samples were centrifuged and stored at –20°C with the addition of sodium metabisulfite until used. The method of our determining plasma norepinephrine levels [11] consisted of purification [3] and gas chromatography-mass fragmentgraphic analysis [4]. It determined norepinephrine contained in the mixture of plasma and deuterated norepinephrine, the latter being added in advance as the internal standard, by elution in boric acid gel column and by susequent gas chromatography-mass fragmentgraphy (Jeol JMS model D-100) after conversion into its pentafluoropropionic derivative.

Results

Group A. The mean plasma norepinephrine level (mean ± SD) in the 23 patients with Ménière's disease and the 12 with sudden deafness, both in supine position was 240 ± 98 and 230 ± 86 pg/ml, respectively. They did not differ significantly in mean plasma norepinephrine level from the controls (260 ± 89 pg/ml).

Figure 1 shows the increments of plasma norepinephrine level and mean arterial pressure at 5 and 10 min after standing upright from a supine position. The two parameters increased significantly in the controls as well as in the patients as follows. In healthy adults, norepinephrine increased from 260 ± 89 to 420 ± 100 pg/ml (5 min) and 470 ± 73 pg/ml (10 min). Mean arterial pressure rose from 81.6 ± 12.2 to 101.0 ± 14.9 mm Hg and 97.6 ± 15.9 mm Hg. In Ménière's disease patients, norepinephrine increased from 240 ± 98 to 430 ± 160 pg/ml and 460 ± 160 pg/ml. Mean arterial pressure rose from 81.6 ± 12.2 to 88.3 ± 14.3 mm Hg and 89.3 ± 16.5 mm Hg. In sudden deafness patients norepinephrine increased from 230 ± 86 to 400 ± 170 pg/ml and 430 ± 180 pg/ml. Mean arterial pressure rose from 86.9 ± 12.4 to 93.9 ± 21.5 mm Hg and 98.6 ± 18.8 mm/Hg. The difference in increment of plasma norepinephrine levels between patients and controls was below the significant level. The increment of mean arterial pressure at 5 and 10 min while the subjects continued to stand upright was smaller in patients than in controls, particularly in patients with Ménière's disease.

The fact that the patients in an upright position showed less elevation of mean arterial pressure than controls in spite of normal norepinephrine release prompted us to compute the ratio of norepinephrine increment to mean arterial pressure increment as an index to the responsiveness of arterial pressure to norepinephrine.

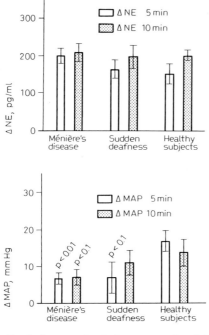

Fig. 1. Increments of plasma norepinephrine level (*a*) and mean arterial pressure (*b*) after standing up for 5 and 10 min.

As a result (fig. 2), patients with Ménière's disease showed the significantly lower ratios than healthy adults in only 5 min after standing upright.

Group B. In the 6 healthy adults, the increment of mean arterial pressure in norepinephrine infusions was proportional to the log of the plasma norepinephrine level (fig. 3). During norepinephrine infusion, plasma norepinephrine levels usually increased in direct proportion to the rate of infusion. The slope of the line determined by linear regression analysis was determined for each subject and taken as the gain. The plasma norepinephrine level at which the increment in mean arterial pressure extrapolated to zero was taken as the threshold level. Table II shows the average values of the gain and the threshold level in plasma norepinephrine concentration in patients with Ménière's disease, sudden deafness and controls. The threshold levels of plasma

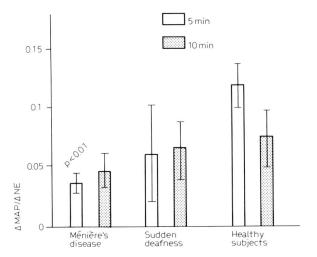

Fig. 2. Ratios of the increments of mean arterial pressure to those of norepinephrine after standing up for 5 and 10 min.

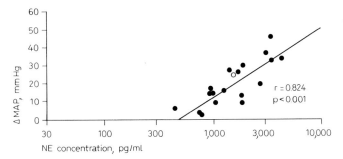

Fig. 3. Relationship of plasma norepinephrine levels to the increment of mean arterial pressure during intravenous infusion of norepinephrine in healthy adults.

norepinephrine were of the same order for both patients and controls. Ménière's disease patients differed from healthy adults and sudden deafness patients in pressor response to infused norepinephrine and showed significantly decreased gain. Moreover, examining the gain in patients and controls according to three divisions of blood pressure, i.e., systolic, diastolic and mean arterial pressure, the decreased gain in Ménière's disease patients was more remarkable in diastolic arterial pressure than in mean arterial pressure and below the significant level in systolic blood pressure (table III).

Table II. Gain and threshold for pressor response to infusion of norepinephrine (mean ± SEM)

	Control n = 6	Ménière's disease n = 10	Sudden deafness n = 5
Gain	45.1 ± 6.0	30.0 ± 3.4*	43.2 ± 10.6
Threshold	540 ± 104	584 ± 73.1	479 ± 121

* p < 0.05.

Table III. Gain for pressor response to infused norepinephrine in systolic, diastolic and mean arterial blood pressure (mean ± SEM)

	Systolic pressure	Diastolic pressure	Mean arterial pressure
Control n = 6	63.1 ± 6.1	34.6 ± 5.4	45.1 ± 6.0
Ménière's disease n = 10	55.2 ± 8.2	17.1 ± 2.1**	30.0 ± 3.4*
Sudden deafness n = 5	83.3 ± 21.7	33.1 ± 12.8	43.2 ± 10.6

*p < 0.05; **p < 0.01.

Discussion

In group A, both Ménière's disease and sudden deafness patients had normal plasma norepinephrine levels in supine position and normally responded to postural change. *Ziegler* et al. [10] classified orthostatic hypotension according to plasma norepinephrine levels after a postural change, i.e. from supine to upright, into type 1 primary autonomic dysfunction and type 2 primary autonomic dysfunction, the latter associated with central nervous system disorders, such as Shy-Drager syndrome and Parkinson's disease. They reported that type 1 dysfunction patients have low plasma norepinephrine levels that do not respond to postural change but type 2 patients have normal plasma norepinephrine levels in supine position and do not respond to postural change. They suggested that type 1 dysfunction patients have an organic defect in the peripheral sympathetic system, whereas type 2 patients have defects in the central sympathetic system.

These findings led us to conclude that patients with Ménière's disease and sudden deafness have no organic defects in the central or peripheral sympathetic system. A finding of particular interest for us was that only Ménière's disease patients showed a significant decrease in ratios of increments in mean arterial pressure to those in norepinephrine levels only after 5 min standing up (fig. 2).

We suggest that the hyporesponsiveness of mean arterial pressure to endogenously released norepinephrine may be due to the impaired sympathetic effector organ in arterial walls. At present the cause of disappearance of hyporesponsiveness after 10 min standing upright remains unknown.

The findings in group A prompted us to conduct another experiment in group B to elucidate whether Ménière's disease patients also show the hyporesponsiveness of mean arterial pressure to exogenously infused norepinephrine. The grade of the pressor response to infused norepinephrine depends generally on the sensitivity of sympathetic effector organ in arterial walls. The reduced sensitivity to the pressor substance decreases the slope of the line determined by linear regression analysis relating increments of blood pressure to the log of plasma norepinephrine levels. Only Ménière's disease patients showed a significantly reduced gain (the slope) as compared with controls. The lowest gain of stimulus-response curve was observed in diastolic phase of the blood pressure (table III). On the other hand, the threshold levels, which norepinephrine yields critically the pressor response, were of the same order among all subjects. These findings imply that patients with Ménière's disease revealed also the hyporesponsiveness of mean arterial pressure, especially of diastolic arterial pressure, to exogenously infused norepinephrine and that their sympathetic effector organs in arterial walls, especially in arteriolar walls, have the subnormal sensitivity to the pressor substance.

From the above-mentioned findings in experiments A and B we can suppose that patients with Ménière's disease differ from patients with sudden deafness in that the responsiveness of blood pressure to endogenous and exogenous norepinephrine was low. The conclusive speculation requires a more precise analysis of plasma norepinephrine

[2], because the hyporesponsiveness is also affected by a variety of factors including synthesis, storage, release, reuptake and turnover of norepinephrine.

Polinsky et al [7] classified orthostatic hypotension according to gain and threshold level of plasma norepinephrine-blood pressure curve relating to exogenously infused norepinephrine into three sorts of patients, i.e. idiopathic orthostatic hypotension (type 1 dysfunction), patients with multisystem atrophy (Shy-Drager syndrome, type 2 dysfunction) and sympathotonic orthostatic hypotension (type 3 dysfunction). Type 1 dysfunction had a shift to the left of the plasma norepinephrine-blood pressure curve, suggesting denervation sensitivity. Type 2 dysfunction shows the increase of the slope in the curve, implying deficient reflex modulation in central autonomic nervous system. Type 3 dysfunction, which is attended by marked tachycardia, revealed the subsensitivity to administered norepinephrine and suggests the deficient responsiveness of sympathetic effector organ. This dysfunction is similar to those in our patients with Menière's disease, because the many patients with hyporesponsiveness also showed the aberrations in Schellong's test characterized by the increased pulse rate.

Lastly present findings will give another useful measure to make differential diagnosis between Menière's disease and sudden deafness with vertigo, and we believe that the hyporesponsiveness of blood pressure to endogenously released and exogenously infused norepinephrine reduces transiently the blood flow to the inner ear and contributes to the onset of Menière's disease.

References

1 Engelman, K.; Portnoy, B; Lovenberg, W.: A sensitive and specific double isotope derivative method for the determination of catecholamines in biological specimens. Am. J. med. Sci. *255:* 259–268 (1968).
2 Euler, M.: Assessment of sympathetic nervous function in humans from noradrenalin plasma kinetics. Clin. sci. *62:* 247–254 (1982).
3 Higa, S.; Suzuki, T.; Hayashi, A.; Tsuge, I.; Yamamura, Y.: Isolation of catecholamines in biological fluids by boric acid gel. Analyt. Biochem. *77:* 18–24 (1977).
4 Karoum, F.; Cattabeni, F.; Costa, E.; Ruthven, C.R.J.; Sandler, H.: Gas chromatographic assay of picomole concentration of biogenic amines. Analyt. Biochem. *47:* 550–561 (1972).
5 Matsunaga, T.; Yoshida, J.; Yoshino, K.: Role of plasma norepinephrine in the diagnosis of Menière's disease, pp. 182–188. (Thieme, Stuttgart 1981).

6   Naito, T.; Sugiyama, S.; Sakai, K.: Experimental studies on Ménière's disease. Proc. Vth Extraordinary Meeting of the Barany Society, 1975, pp. 32–35.

7   Polinsky, R.J.; Kopin, I.J.; Ebert, M.H.; Weise, V.: Pharmacologic distinction of different orthostatic hypotension syndromes. Neurology *31:* 1–7 (1981).

8   Schuknecht, H.F.: Pathology of the ear, pp. 453–465. (Harvard University Press, Cambridge 1974).

9   Williams, H.L., Jr.: A review of the literature as to the physiologic dysfunction of Ménière's disease. A new hypothesis as to its fundamental cause. Laryngo *75:* 1661–1689 (1965).

10   Ziegler, M.G.; Lake, C.R.; Kopin, I.J.: The sympathetic nervous system defect in primary orthostatic hypotension, New Engl. J. Med. *296:* 293–297 (1977).

11   Yoshida, J-I.; Yoshino, K.; Matsunaga, T.; Higa, S.; Suzuki, T.; Hayashi, A.; Yamamura, Y.: An improved method for determination of plasma norepinephrine: isolation by boric acid gel and assay by selected ion monitoring. Biomed. Mass Spectrom. *7:* 396–398 (1980).

T. Matsunaga, MD, Osaka University Medical School, Osaka (Japan)

Adv. Oto-Rhino-Laryng., vol. 30, pp. 254–257 (Karger, Basel 1983)

Autoimmune Sensorineural Hearing Loss as an Aggravating Factor in Ménière's Disease

John J. Shea

Shea Clinic, Memphis, Tenn., USA

The exact role of autoimmune response in otherwise typical Ménière's disease is difficult to determine. We have come to suspect that on one side you can have typical Ménière's disease, due to an inadequate endolymphatic sac, the classical explanation, and on the other you have typical autoimmune sensorineural hearing loss, due to immune-mediated vasculitis, as described by *McCabe* [1], which can resemble rapidly progressive classical Ménière's disease. In between you have patients with an inadequate endolymphatic sac in which autoimmune response occurs in much greater frequency than it does in the general population, as judged by the favorable response to dexamethasone in about 10% of patients with otherwise typical Ménière's disease. Why patients with an inadequate endolymphatic sac, and the other predisposing factors to Ménière's disease, have a greater amount of autoimmune response than is found in the general population is difficult to explain, but it must be that something in the developing pathology of Ménière's disease triggers the autoimmune response. The tests for autoimmune response such as the lymphocyte migration inhibition test using inner ear tissue as antigen as recommended by *McCabe* [1] and the monocyte migration inhibition test using type II collagen as the antigen are not reliable, since only about half of clinically obvious autoimmune sensorineural hearing loss patients will have a positive result. For that reason we have stopped performing these tests and rely on the

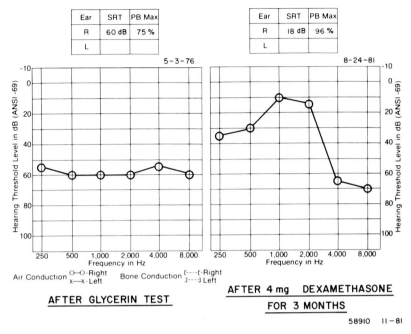

Fig. 1. Improvement in hearing after 4 mg dexamethasone daily for 3 months.

clinical picture to make the presumptive diagnosis of autoimmune disease, and hope to confirm the diagnosis by the response to moderately large doses of dexamethasone.

We came to this conclusion in a roundabout way. In April 1976, I first saw a 32-year-old young man with a long history of typical bilateral Ménière's disease. He had no hearing in one ear, after transmastoid labyrinthectomy, and a severe non-fluctuating hearing loss in the other ear after an epidural shunt with Gelfilm® in the sac. To my great surprise he had an immediate dramatic improvement in hearing after taking 4 mg dexamethasone for only a few days (fig. 1).

This patient was 5 feet tall and weighed only 140 pounds because of familial vitamin D-resistant rickets, for which he had several operations to straighten and shorten his legs. It may be that this metabolic disorder may account for part of his remarkable response to dexamethasone.

He had an unusually good response to glycerin on three occasions and a prompt drop in hearing when given intravenous histamine.

Monocyte migration inhibition test using purified type II collagen as found in the otic capsule as the antigen showed no inhibition. Because of the prompt decline in hearing following intravenous histamine, the hearing loss was suspected of being due to or aggravated by an autoimmune response. The patient was given a moderately large dose of dexamethasone, 4 mg daily, for 3 months, upon which he made a dramatic improvement in hearing, from 63 dB average for the three speech frequencies with 72% discrimination, to 22 dB average with 88% discrimination and complete relief from repeated attacks of vertigo. The patient had typical elevated negative summating potential of Ménière's disease in his transcanal electrocochleogram before and after dexamethasone treatment, and spontaneous right-beating nystagmus with reduced caloric response in both ears.

We have many other patients with unilateral and bilateral hearing loss due to typical Ménière's disease who recovered hearing, but less dramatically, with this rather large dose of dexamethasone. About 10% of all patients with typical Ménière's disease will show hearing gain and/or dramatic reduction in dizziness with dexamethasone treatment. We are now routinely giving all patients with rapidly progressive fluctuant hearing loss and attacks of vertigo with the diagnosis of Ménière's disease large doses of dexamethasone, 4 mg daily for 10 days and 0.75 mg every other day for 20–80 days. We believe autoimmune response to be a possible trigger to tip the balance of compensation in patients with anatomically inadequate sacs with Ménière's disease.

Just what dexamethasone and the other steroids do in these patients with typical autoimmune sensorineural hearing loss and/or those with Ménière's disease aggravated by autoimmune response is difficult to say. In the dose and for the duration given steroids such as dexamethasone are known to decrease circulating lymphocytes and reduce immune globulin in the blood. Just how these steroids have the effect they do in the inner ear is not known. One of our first patients with a clinical picture more suggestive of autoimmune sensorineural hearing loss than Menière's disease made a good recovery of hearing in her one remaining ear with large doses of dexamethasone, after this hearing was depressed by intravenous histamine. Unfortunately on this dose she developed such severe symptoms of steroid intoxication that she had to stop taking dexamethasone and her hearing went down to its former level. We were able to get her hearing back to its best level with a small dose of dexamethasone plus weekly plasmopheresis to remove the very large amount of abnormal globulin in her circulating blood. Presumably this plasmopheresis/low-dose steroid regimen could be used on many patients with autoimmune sensorineural hearing loss to reduce the amount of steroid required, even when given every other day.

Reference

1   McCabe, B.F.: Autoimmune sensorineural hearing loss. Ann. Otol. Rhinol. Lar. *88*: 585–589 (1979).

J.J. Shea, MD, Shea Clinic, 1080 Madison Avenue, Memphis, TN 38104 (USA)

Adv. Oto-Rhino-Laryng., vol. 30, pp. 258–263 (Karger, Basel 1983)

Ultrafine Structure of the Otoconial Membrane

Y. Harada

Department of Otolaryngology, Hiroshima University School of Medicine, Hiroshima, Japan

Introduction

It is already known that the otolithic organ has numerous small otoconial crystals on the otoconial membrane. Recent electronmicroscopic studies provided some knowledge about the morphology and composition of otoconia [1–3]. Basically mammalian otoconium is composed of calcium carbonate and organic matrix [4, 5]. Concerning the genesis of otoconia, some investigators reported that in lower vertebrates otoconia are originated from the endolymphatic duct or sac and also from the labyrinth [6, 7]. In mammals, however, it is expected that the origin of otoconia might differ from that of lower vertebrates. Recent studies on the process of otoconial growth revealed that otoconia are formed by gradual deposition of calcium carbonate around the nucleus [4, 5, 8]. However, the present study suggested somewhat different results about otoconial formation. A couple of years ago the author observed the otoconial membrane and the surface of sensory epithelium, using SEM and X-ray microanalyzer, and was under the impression that the calcium of otoconia is secreted from the supporting cells. In the present study, the otoconial membrane and sensory epithelia of the guinea pig were observed in more detail and a new theory about the mechanism of otoconial formation was obtained.

Materials and Methods

The subjects were 10 guinea pigs (non-albinos about 300 g body weight) with normal pinna reflex. The temporal bones were quickly dissected and immersed in 2% OsO_4 for 1 h. The specimens were dehydrated in ethanol, placed in isoamylacetate and critical-

Fig. 3. A part of the otoconial membrane which is detached near the undersurface. There are numerous globular substances. Some of them are transforming into the otoconial crystal (arrows).

Fig. 4. The surface of neurosensory epithelia. Some protusions from the supporting cells can be seen (arrows).

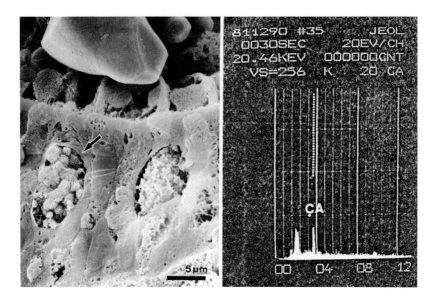

Fig. 5. a Cracked surface of neurosensory epithelia. Small granular substances are observed inside a cell which is considered to be a supporting cell (arrow). b X-ray microanalysis of calcium contain of the small granular substances.

cracked specimens, there were some cells possessed small granules which are from 1.0 to 2.0 μm in diameter (fig. 5). These granules also contained about as much calcium as the globular substance. In conclusion, a precursor of the otoconium, which was observed as a globular substance, is secreted from the supporting cell. Our findings suggest that immature otoconia are metamorphosed into mature types due to concentration of calcium, and do not support the previous theory which is based on the seeding of the nucleus and the following deposition of calcium.

References

1   Carlström, D.; Engström, H.; Hjorth, S.: Electron microscopic and X-ray diffraction studies of statoconia. Laryngoscope 63: 1052–1057 (1953).
2   Carlström, D.; Engström, H.: The ultrastructure of statoconia. Acta oto-lar. 45: 14–18 (1955).

3   Sanchez-Fernandez, J.M.; Marco, J.; Rivera-Pomar, J.M.; Delgado, R.M.: Electron diffraction studies on otolithic organization in the macula utriculi of the guinea pig. Acta oto-lar. *73:* 267–269 (1972).
4   Lim, D.J.: Formation and fate of the otoconia. Ann Otol. Rhinol. Lar. *82:* 23–25 (1973).
5   Ross, M.D.; Johnsson, L.G.; Peacor, D.; Allard, L.F.: Observations on normal and degenerating human otoconia. Ann Otol. Rhinol. Lar. *85:* 310–326 (1976).
6   Vilstrup, T.: On the formation of the otoliths. Ann Otol. Rhinol. Lar. *60:* 974–981 (1951).
7   Balsamo, G.; Devincentiis, M.; Marmo, F.: The effect of tetracycline on the processes of calcification of the otoliths in the developing chick embryo. J. Embryol. exp. Morph. *22:* 327–332 (1969).
8   Salamat, M.S.; Ross, M.D.; Peacor, D.R.: Otoconial formation in the fetal rat. Ann Otol. Rhinol. Lar. *89:* 229–238 (1980).

Y. Harada, MD, Department of Otolaryngology, Hiroshima University School of Medicine, Kasumi-cho, 1-2-3, Hiroshima 734 (Japan)

Adv. Oto-Rhino-Laryng., vol. 30, pp. 264–267 (Karger, Basel 1983)

An Experimental Investigation of Streptomycin Ototoxicity to the Otolith Organ by the Parallel Swing and Observation of Otoconia through SEM in the Guinea Pig

Takashi Futaki, Isuzu Kawabata

Department of Otolaryngology, University of Tokyo, Hongo, Bunkyo-ku, Tokyo

Introduction

The parallel swing (PS) has been understood as a test for the otolith organ by its alternating linear acceleration since *Wojatschek* [1936], *Jongkees and Groen* [1946] and *Jongkees and Philipszoon* [1963]. Several experimental reports for streptomycin (SM) 'ototoxicity' [*DeKleyn and van Deinse,* 1950; *McCabe,* 1964] have revealed the stronger toxic effect to the hair cell of the canal than that to the otolith organ. *Harada and Sugimoto* [1977] presented the fate of otoconia including the influence of SM intoxication through SEM.

The authors wish to verify and to compare the toxic effects of SM to the otolith organ in both the physiological and morphological aspects simultaneously in the guinea pig.

Methods

24 white guinea pigs (250–300 g b.w.) were divided into five groups, i.e. one control group which was injected saline and two groups which were administrated SM intramuscularly in a dosis of 250 mg/kg every 2 days, seven times and adding methyl-B_{12} (500 μg) twice every day. Prior to administration the PS was performed in all animals and the sinusoidal eye movement was recorded electronystagmographically using the leg-free cramp through the needle electrodes. After 2 and 4 weeks the final injection – the same as before – the PS was carried out and the guinea pigs immediately sacrificed for SEM observation of otoconia.

Table I. Results of the PS in five groups (SD)[1] – Sa/Sb –

	n	Amplitude	Phase
Control	4	1.095 (0.473)	0.920 (0.189)
S-1	4	0.455 (0.379)	1.583 (0.805)
S-2	5	0.475 (0.202)	1.167 (0.182)
SM-1	5	0.7557 (0.1147)	1.6510 (0.5206)
SM-2	5	1.0737 (0.4993)	2.3040 (0.9446)

[1] Sa/Sb = the ratio of the group values before and after feeding; S-1 = the one SM administration (250 mg/kg, 1/2 days, 7 times) only with after-feeding of 2 weeks; S-2 = the same one with that of 4 weeks; SM-1 = addition of methyl-B_{12} (500 µg/kg, 1/1 days, 14 times) to S-1; SM-2 = the same as to S-2 (cf. fig. 1).

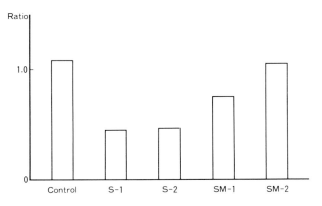

Fig. 1. The changes of amplitude in PS (cf. table I).

Results

Concerning the amplitude, the ratio of the average values before and after the 2 weeks feeding with injected saline in the control groups yielded 1.09 which showed no remarkable change. On the other hand, the ratio of amplitude in SM-administered guinea pigs presented values of 0.45 and 0.47 which resulted as a significant reduction (table I). This reduction means the depressed excitability of the otolith organ by intoxication. In the two groups where methyl-B_{12} was added, the reduction was as half as SM alone and recovered in one group (fig. 1). The phase lag ratio was significantly greater than that of the control group.

Fig. 2. SEM of the otolith membrane (left, low magnification) and the otoconia (right, high magnification) of the guinea pig after saline injection; the control groups.

Fig. 3. SM-administered group (cf. fig. 2).

In the SEM investigation of otoconia, the control group showed that multiple layers and multiple sizes of the otoconial membrane consisted of regular, hexagonal-shaped crystals with sharp cut-ends and smooth surfaces (fig. 2). In SM groups, the survey of the membrane resulted in thin layers, with small-sized crystals without any giant stones. The features of each otoconia were irregular in shape as a crystal, not hexagonal but cylindrical with a rough surface and blunt edge (fig. 3). In groups where methyl-B_{12} was added, the surface view of the otolith membrane looked like an intermediate grade of the former two.

Discussion

It would be conceivable that the canal system receives stronger damage than the otolith organ. With regard to the amplitude of the sinusoidal eye movement brought about by PS, the SM groups yielded a half of the significant reduction compared to that of the control group. This reduction of response could be considered as the evidence of the ototoxic effect of SM on the otolith organ of the guinea pig. In simultaneous observations of the otolith membrane by SEM, the morphological changes have been illustrated in the SM groups, i.e. the thin layer without multiplicity in their size and the irregular otoconia, not hexagonal with a rough surface and blunt edges resembling that under zero g. [*Vinnikov* et al., 1979]. These features may suggest some damage by SM in turning over of the otolith production and absorption, and might be closely linked to the reduction of the amplitude of the eye movement induced by PS.

References

DeKleyn, A.P.H.A.; Deinse, J.B. van: The influence of streptomycin on the vestibular system. Acta oto-lar. *38:* 3–7 (1950).

Harada, Y.; Sugimoto, Y.: Metabolic disorder of otoconia after streptomycin intoxication. Acta oto-lar. *84:* 65–71 (1977).

Jongkees, L.B.W.; Groen, J.J.: The nature of the vestibular system. J. Laryng. *61:* 529–541 (1946).

Jongkees, L.B.W.; Philipszoon, A.J.: The influence of position upon the eye movements provoked by linear accelerations. Acta oto-lar. *56:* 414–420 (1963).

McCabe, B.F.: Nystagmus response of the otolith organs. Laryngoscope *74:* 372–381 (1964).

Vinnikov, Y.A., et al.: The structural and functional organization of the vestibular apparatus of rats exposed to weightlessness for 20 days on board the sputnik 'Kosmos-782'. Acta oto-lar. *87:* 90–96 (1979).

Wojatschek, W.: Klinische Messung der Otolithenfunktion. Acta oto-lar. *234:* 11–33 (1936).

T. Futaki, MD, Department of Otolaryngology, University of Tokyo, Hongo, Bunkyo-ku, Tokyo (Japan)

Adv. Oto-Rhino-Laryng., vol. 30, pp. 268–270 (Karger, Basel 1983)

Diagnostic Significance of
Neck Vibration for the Cervical Vertigo

Hideyuki Iwasa, Toshiaki Yagi, Tomokazu Kamio

Nippon Medical School, Department of Otolaryngology, Sendagi, Bunkyo-ku, Tokyo, Japan

Introduction

Since *Magnus* reported on postural change brought about by neck torsion, it has been generally accepted that the neck afferents play a major role in postural adjustment. However, since a method for the diagnosis of dizziness caused by abnormalities of the neck reflexes has not been established yet, there is a possibility that they are being diagnosed as other diseases or a disease of unknown origin. In this study, eye and body movements caused by vibratory stimulation of the dorsal neck area were examined in patients with vertigo. The efficacy of vibration as an objective examination of dizziness caused by abnormalities in the neck proprioceptors was also investigated.

Method

Subjects. The subjects were 55 patients with unilateral peripheral vestibular impairment, and 58 patients with vertigo, possibly of cervical origin. Unilateral peripheral vestibular impairment means unilateral labyrinthine dysfunction verified with a caloric test and audiological examination, but without any findings of central nervous system dysfunction. The cases of suspected cervical vertigo were patients who had nuchal pain or neck stiffness accompanied by dizziness and sympathetically provoked symptoms such as nausea, palpitation, perspiration or blurred vision, but without findings suggesting central nervous system dysfunction or peripheral vestibular dysfunction. As a control, 30 healthy adults in their twenties and not suffering from dizziness were tested with the same methods.

	Cervical vertigo (n = 58)	Unilateral vestibular impairment (n = 55)	Normal (n = 30)
Nyst. (−)	19 (23%)	14 (25%)	25 (83%)
Nyst. (+)	39 (67%)	41 (75%)	5 (17%)
	[→][→]/[←][←] 9	[←][←] (contra.) 41	[→][←] 2
	[→][←] 16	[→][→] (ipsi.) 0	[←][→] 2
	[←][→] 14	[←][→]/[→] 0	[←] 1

[ ][ ]
r-st. l-st.

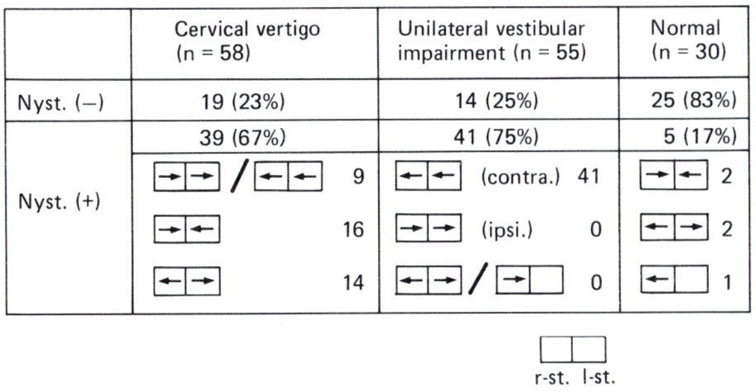

Fig. 1. Direction of neck-induced nystagmus.

Procedure. A specially designed vibrator with a frequency of 125 Hz was used for the vibratory stimulation of the neck. Eye movement was observed behind Frenzel's glasses with ENGs being recorded simultaneously. To observe body movement induced by the neck stimulation, the same apparatus was attached to the back of the subjects' neck with a belt. The posturogram for the center of gravity was recorded on an X-Y recorder. Moreover, in order to observe the effect of neck stimulation on the vestibulo-ocular reflex, vibration was applied to the back of the neck while the subjects exhibited caloric nystagmus.

Results

In a few of the 30 healthy adults, fine horizontal nystagmus and postural instability resulted, but the direction of the nystagmus and the postural deviation were not constant (fig.1). In 41 (74.5%) of the 55 cases with unilateral vestibular impairment, nystagmus was induced by the neck stimulation. In 29 of these 41 cases, spontaneous nystagmus was observed in positional or positioning nystagmus tests. In all the 41 cases the neck-induced nystagmus was horizontal with a rotatory component and always contralateral to the impaired inner ear. Postural deviation ipsilateral to the impaired labyrinth, in other words, on the same side as the slow phase of the neck-induced nystagmus, was often observed when both sides of the neck were stimulated.

Of the 58 cases with suspected cervical vertigo, the neck stimulation produced nystagmus in 39 cases (67.2%).In 9 of these, nystagmus on one side was produced by stimulation of either the right or left side of the neck. In 16 cases, nystagmus was produced contralaterally to the stimulated side. In the other 14 cases, nystagmus was induced on the same side as the stimulation. In 29 of these 39 cases, there was no positional or positioning nystagmus. Although the type of body movement induced by neck stimulation in the cases of cervical vertigo varied, there was a tendency to deviate on the contralateral side of the evoked nystagmus. In cases which had localized tender spots or bruises on the neck, violent dysequilibrium was often provoked by the vibratory stimulation of these points. Caloric nystagmus was often influenced by neck stimulation. Generally, if the direction of caloric nystagmus coincided with that of neck-induced nystagmus, the caloric nystagmus was reinforced and vice versa.

48 patients with vertigo of unknown origin because of no neurotological abnormalities were examined to discover if the vertigo would have been caused by neck disorders. In 25 cases (52.1%), nystagmus was induced by the neck stimulation. The patients with the neck-induced nystagmus had the following clinical characteristics when compared to the patients without neck-induced nystagmus; (1) the ages of the patients were higher; (2) roentgenological abnormalities in the neck were seen more often; (3) accompanying symptoms of dizziness were often neck stiffness or sympathetically provoked symptoms, such as nausea, palpitation or blurred vision. In 14 (56%) of the 25 cases who had neck-induced nystagmus, changing the side of stimulation reversed the direction of the nystagmus.

Conclusion

The usefulness of neck vibration in the objective diagnosis of cervical vertigo and in differentiating between cervical vertigo and vertigo with peripheral vestibular impairment was confirmed. Moreover, we indicated the possibility that in the cases with vertigo of uncertain cause, there may be cases of vertigo originating from the abnormality of the neck proprioceptors, if they are examined precisely by the neck vibration.

H. Iwasa, MD, Nippon Medical School, Department of Otolaryngology,
Sendagi, Bunkyo-ku, Tokyo 113 (Japan)

Adv. Oto-Rhino-Laryng., vol. 30, pp. 271–273 (Karger, Basel 1983)

Vestibuloneurological Findings in Children with Severe or Profound Loss of Hearing

Hana Krejčová, Ivan Lesný, Marie Vacková, Blazena Stračárová

Charles University, Prague, Czechoslovakia

A group of 30 preschool and school children of the age of 3–13 years was examined. All the children underwent ENG and EEG examination, conventional neurologic examination, special examination for hypotony, praxis and gnosis, and the primitive locomotion test. Audiological examination revealed in all the children severe hearing losses of 70 and more dB (measured in the speach range 500, 1,000 and 2,000 Hz). Residual hearing was in some children restricted to a small range reaching to 500–1,000 Hz. In most children the whole range up to 3,000–4,000 Hz was covered. All the children were tested by nonverbal psychological tests and children with suspected mental deficiency were not included. A hereditary conditioned hearing impairment was found in 29% of the children, congenital hearing loss in 58%, and postnatal causes of hearing defects of various etiology was evident in 12% of children.

None of the children revealed spontaneous nystagmus, gaze nystagmus was present in 3% and direction-fixed positional nystagmus in 51% of the cases. Caloric and rotational hyperreflexia was present in 25%, caloric hyperreflexia in 12%, rotational in 6% of cases. Caloric and rotational areflexia was found in 12%, in 3% there was only caloric but not rotational areflexia. Directional preponderance was present in 22%, in 10% of them of central and in 12% of peripheral origin. An impaired optokinetic response was present in 32% – in 10% we found bilaterally irregular optokinetic nystagmus, in 22% the asymmetry was caused by directional preponderance. Eye tracking test (ETT) was impaired in 9% of the subjects.

The evidence of central vestibular impairment was found in 35%, peripheral vestibular impairment in 26% and an ambivalent type of vestibular lesion in 16% of the examined children. In 25% of the children the conventional neurological examination revealed the syndrome of the so-called 'minor brain disturbance', e.g. altered reflexes, pyramidal signs, dyslexia etc. Central hypotony was found in 70% of the tested children. All signs of hypotony were significantly positive in comparison with normal children. In 51% of the children, disturbances of the primitive locomotion test were found. Dyspraxia-dysgnosia was proved in 28% of the examined children. The central hypotony was found very similar to the findings typical for the hypotonic form of central palsy or hypotonies often found in blind children. Presumably, a great part of central hypotonies which in later childhood do not change into other syndromes are caused by lack of afferent stimuli.

The majority of children had abnormal EEG findings. Three types of EEG abnormalities were observed: slight generalized slowing in 22%, bisynchronous slow waves, mostly bioccipital slow synchrony in 25%, and in 32% of cases the most frequent abnormality – less mature EEG.

The multifactorial causes of severe loss of hearing and profound deafness are accompanied by various vestibular impairments in which there is no significant correlation between the degree of hearing and vestibular disorder. Only in vestibular areflexia was most advanced hearing loss observed. The central vestibular syndromes found in 33% of children were not influenced by the degree of hearing involvement but by the impairment of CNS particularly by the cerebellum which presumably caused the disinhibition of vestibular reactions.

Conclusions

Vestibular findings in children with severe and profound hearing loss are not dependent upon the degree of hearing impairment. The only exceptions are cases with bilateral vestibular areflexia in whom the most profound hearing losses were proved. In 94% of the tested children, neurological impairments of various types were found, particularly the central hypotonic syndrome. In 81% of the cases, abnormal EEG recordings were repeatedly found. It seems that hearing loss is accompanied by retarded CNS development, particularly of the cerebellum. The symptoms of persistent central hypotony, dyspraxia-dysgnosia as well as the abnormalities found in the EEG recordings are the typical signs of a less-mature brain.

H. Krejčová, MD, Charles University in Prague, Neurological Department, Phoniatric Department, Prague 5-Motol (Czechoslovakia)

Adv. Oto-Rhino-Laryng., vol. 30, pp. 274–277 (Karger, Basel 1983)

Moving Sound Sources and Their Clinical Applicability

K.-P. Schaefer, K.-J. Suess, H.A. Friedrich

Neurobiology Unit, Department of Psychiatry, Göttingen, FRG

The auditory organ seems to insure an acoustic spatial stability, which, due to autonomic and environmental conditions, make possible the acoustic orientation and the reliable location of sound sources or acoustic targets in the surroundings. However, only limited detailed information is available regarding the form and occurrence of these acoustically induced orientation reactions and what is known has not been considered under clinical conditions [5].

In routine clinical examinations as audiometry and auditory evoked responses the acoustic signals are generally applied via earphones. The localization of sound in space and the perception of audiokinetic phenomena, however, are only possible under natural conditions taking into consideration the acoustic shadow of the head, the resonance conditions of the auricles, the outer auditory canal, etc. Therefore, the arrangement, which we prefer to use for laboratory application of moving acoustic signals, consists of 16 loudspeakers set up in a circle around the test subject. In order to achieve the effect of continuous movement of the sound source between two loudspeakers the intensity of one loudspeaker is increased and accordingly decreased at the adjacent channel while the total intensity is kept constant (stereophonic listening, movement of phantom) [1].

With the aid of the above-mentioned method we now have started to investigate more closely the influence of moving acoustic signals on the cortical, subcortical and the spinal motor functions in the human being. Following are descriptions of three examples for application which may also possess clinical significance.

In general, the cortically-evoked potential is acoustically stimulated by a standard stimulus with 150 ms duration and definitive on-off limitation. Corresponding studies on moving acoustic signals have not been published yet.

If under the described conditions – not applied via headphones – a standard stimulus (70 individual signals) is stationary in the anterior median, the previously described typical vertex potential can also be seen. However, if the signal is moved, the evoked potential loses some clarity. In further investigations, a continuously applied steady sound signal was moved from right to left and back at a frequency of 0.5–0.25 Hz, the originally described vertex potential is observed once again [6].

In another series of experiments the influence of moving acoustic signals upon body sway of normal standing man has been investigated. Placed upon a measuring platform 75% of all the test subjects showed phase synchronous body sway with circular movements of the sound signal at frequencies of less than 1.0 Hz. By means of the averaging technique these phenomena can be computed from a total of 50 sound circles. However, using the Fourier transformation no prominent peak can be seen at the fundamental sound movement frequencies or higher harmonics. 90% of the power was below 0.48 Hz, corresponding to other investigations with spontaneous body sway and those under visual stimulating conditions [6].

Eye movements, for example optokinetic and vestibular nystagmus, are used as essential diagnostic criteria for peripheral and central vestibular pathology. However, little is known about eye movements as the possible indicator of central auditory disturbances. In the present article, voluntary and involuntary tracking eye movements in response to circular movements of the sound signal in various spatial planes (frame of reference relative to the head) are described.

The investigations in the horizontal plane confirm the findings from earlier investigations. Voluntary and involuntary tracking eye movements appear in the nystagmogram as so-called 'staircase jerks' with saccades of different size and frequency. Continuous tracking eye movements which correspond to the visual tracking eye movements upon moving points of light (smooth movements), with the exception of single, possible pathological cases, have not been observed.

With investigations in the frontal plane the physical situation basically corresponds to the situation of rotating movements in the horizontal plane, i.e. the most important parameters for directional hearing, the difference in running time and the difference in intensity can be utilized for localization purposes by the acoustic organ. Accordingly the investigations have arrived at basically the same results. However, computation of the horizontal and the vertical nystagmogram has frequently resulted in a reduction of the overall amplitude and of the beating field in the vertical direction, especially in darkness, at higher rotation frequency and with a fatigued test subject. In individual cases these findings have been most distinctly with involuntary tracking eye movements [2, 3, 4].

The results from our own investigations with circular movements in the median plane, to begin with, are not as expected. The results obtained by other authors show that the localization of the sound in the median plane is rather imprecise. In the present investigations, however, the test subjects were capable of tracking the circular sound movements with sufficient phase synchronization. According to number and amplitude of saccades the computed median values of 16 test subjects correspond largely to those values obtained with horizontal circular movements. In individual test subjects, however, considerable differences could be observed in comparison with horizontal tracking movements. Not infrequently, the vertical extension of the beating field is greatly reduced, especially in darkness and with regard to the involuntary tracking movements. The tracking eye movements become even more inaccurate when, instead of circulatory movements, low frequency sinusoidal movements of the sound source are performed in the range of the frontal median plane.

On the whole, however, the acoustic system is capable of localizing circular movements in the median plane correctly to the extent that at least the phase relationship is maintained. Since, under these conditions, changes in the sound position do not cause any interaural time or intensity differences, it is obvious that the effect of 'shadowing' and reflection of the auricles, the head and the body must be of considerable importance. Moreover, the present investigation suggests that the additional information of movement or change of position of sound signal seems to considerably improve the performance of discrimination in the median plane. The acoustic system is thus capable of utilizing the differential effect of a change in position of the acoustic signal and to integrate this phenomenon into the processing of the total acoustic information.

References

1   Blauert, J.: Räumliches Hören (Stuttgart 1974).
2   Fiebig, E.; Schaefer, K.-P.; Suess, K.-J.: Form and accuracy of voluntary ocular tracking movements in response to sinusoidally moving acoustic targets. J. Neurol. 226: 77–84 (1981).
3   Fiebig, E.; Schaefer, K.-P.; Suess, K.-J.: Voluntary ocular tracking movements in response to a sinusoidally moving intermittently active sound source. Human Neurobiology 1: 141–144 (1982).
4   Schaefer, K.-P.; Suess, K.-J.: Eye movements induced by acoustic stimuli. EEG – EMG 8: 226 (1977).
5   Schaefer, K.-P.; Suess, K.-J.; Fiebig, E.: Acoustic induced eye-movements. Ann. N.Y. Acad. Sci. 374: 674–688 (1981).
6   Suess, K.-J.; Schaefer, K.-P.; Neetz, A.: Moving sound sources; techniques and applications. Acta oto-lar 94: 29–35 (1982).

Prof. K.-P. Schaefer, Neurobiology Unit, Department of Psychiatry,
v.-Siebold-Strasse 5, D-3400 Göttingen (FRG)

Adv. Oto-Rhino-Laryng., vol. 30, pp. 278–280 (Karger, Basel 1983)

Some Aspects of the So-Called 'Vestibular Neuronitis'

Pekka Silvoniemi, Eero Aantaa

Department of Oto-Rhino-Laryngology, University Hospital of Turku, Finland

Introduction

Vestibular neuronitis is characterized by a sudden vigorous rotating vertigo with a loss of caloric response on the affected side and spontaneous horizontal or rotating nystagmus to the healthy side. There are no symptoms from the other cranial nerves. Patients are young or middle-aged and they have usually no previous case history of vertigo. Often there has been virus infection of upper respiratory airways before, i.e. sinusitis, bronchitis, rhinitis or pharyngitis. The sensation of vertigo diminishes in a few days and patients are usually able to work some weeks after the acute phase of the disease. Recently, mild EEG findings have been reported in neuronitis vestibularis patients [*Anttinen* et al., 1982].

Materials and Methods

For this study 35 patients (18 females, 17 males, mean age 39 years) who had typical case histories of vestibular neuronitis and no caloric reaction on the affected side, were studied again 1 year after the acute phase. No patient had any other disease or medication which could interfere with the results. There were 9 patients suffering from high blood pressure, 4 who had some signs of atherosclerosis, 1 with latent diabetes and 1 with hypothyreosis. In 18 patients there was some kind of infection in the case history.

Pendular eye tracking test (PETT), optokinetic nystagmus (OKN), saccadic eye movements, spontaneous and positional nystagmus and caloric tests with water of 44 and 30 °C were recorded by electronystagmography (Elema Mingograph, direct ink writing, the upper limit being 70 and TC 5 s).

In brainstem-evoked response audiometry (BRA) recordings 60 dB HL 10 Hz condensation/rarefaction clicks were used as stimuli. The lateral electrode was placed on the ipsilateral mastoid process and the midline electrode frontally, at the hairline. The band width of the pre-amplifier was from 20 Hz to 5 kHz/3 dB, 2,000 sweeps were averaged. Reference values obtained from the literature were used in the normative analysis of the recordings for interpeak interval durations and their asymmetry. The results were classified as normal, borderline and pathologic.

Results

The ORL findings of the patients were normal. The pure tone audiograms were normal in all the patients. The speech discrimination scores were normal, too. There was spontaneous nystagmus after PETT recording in 21 patients in the acute phase and in 14 a year after the acute phase. Saccadic eye movements were normal. 31 patients had spontaneous nystagmus after OKN recording and 15 patients had the same result in the follow-up study. Spontaneous nystagmus (mean velocity 21.7°/s) was a constant finding during the acute phase and in 14 patients in the follow-up study (mean velocity 10.5°/s). There were normal caloric findings in 17 patients (49%), remarkably lowered in 6 (17%) and 12 patients (34%) had no caloric reaction on the diseased side at all in the follow-up study. BRA findings were normal in 23 (66%) cases, borderline in 7 (20%) and pathologic in 5 (14%). In borderline cases 1 patient had III–V, 2 had I–III and III–V latency abnormality and 3 patients had very low responses. In patients whose BRA was classified pathologic, abnormality in interpeak latencies were found between I–III in 1 case, between I–III and III–V in 2 cases and between I–V in 2 cases.

Discussion

Facial nerve paralysis, optic neuritis and vestibular neuronitis are to some extent similar diseases of one cranial nerve. The conventional view of vestibular neuronitis is or has been until recently a disease without any central symptoms or without symptoms of other cranial nerves. On the other hand there is material which has shown a lot of findings of central lesions in ENG in these patients [*Wennmo and Pyykkö*, 1982]. Patients have been described to have a vestibular neuronitis associated with facial nerve paralysis and in 1 case a transient abducens paresis [*Depondt*, 1973]. In those materials vestibular neuronitis criteria differed from ours where all the patients had a total loss of caloric function on the affected side. Recently, there have been descriptions of mostly transient EEG abnormalities in vestibular neuronitis patients [*Anttinen* et al., 1982]. In our follow-up material we could find pathologic BRA recordings in 14% of patients. We have got the impression that vestibular neuronitis is seldom a pure lesion of one cranial nerve like facial nerve paralysis or optic neuritis, but there may be a subclinical involvement of brainstem structures. The functional prognosis of vestibular neuronitis is, however, better than in facial paralysis or optic neuritis with which it somehow can be compared.

References

Anttinen, A.; Lang, H.; Aantaa, E.; Marttila, R.: Vestibular neuronitis – a neurological and neurophysiological evaluation (in press).
Depondt, M.: La neuronite vestibulaire. Paralysie vestibulaire à caractères particuliers. Acta oto-rhino-lar. belg. *1973:* 323–359.
Wennmo, C.; Pyykkö, I.: Vestibular neuronitis, a clinical and electrooculographic analysis. Acta oto-lar. (in press).

P. Silvoniemi, MD, Department of Oto-Rhino-Laryngology, University Hospital of Turku, SF-20500 Turku 50 (Finland)

Adv. Oto-Rhino-Laryng., vol. 30, pp. 281–284 (Karger, Basel 1983)

Histopathological Temporal Bone Findings in von Recklinghausen's Disease

A. Bustamante-Balcarcel[a], *R. Barrios del Valle*[b]

[a] Department of Otolaryngology and [b] Department of Pathology,
Instituto Nacional de Cardiologia 'Ignacio Chávez', México, DF Mexico

Biological behavior of neurofibromas in von Recklinghausen's disease (VRD) is very different from that seen in solitary neurinoma of the eight cranial nerves. In this presentation we will analyze some characteristics of the disease in the temporal bone. We studied six temporal bones from 3 patients who suffered from bilateral deterioration of hearing due to VRD. In solitary neurofibroma of the VIIIth nerve, the tumor arises from the superior vestibular branch and less frequently from the inferior branch. In VRD it is possible to find multiple neurofibromas which arise from vestibular nerves or from the cochlear branch in the internal auditory canal. This characteristic was emphasized by *Nager* [1969] (fig. 1). In patients with bilateral neurinomas which are considered incomplete forms of VRD as well as in patients with the full-blown picture, it is possible to find invasion of the scala tympani (fig. 2). They may displace the organ of Corti due to progressive enlargement (fig. 3). These tumors, at least in our cases, were independent and had no relation with the tumor of the VIIIth nerve in the internal auditory canal. *Hallpike* [1963] considered them 'seedlings'. Other authors think that they are manifestations of the multicentric character of these tumors. It is possible to find neurofibromas in the semicircular ducts which arise in the cristae in a way that can occlude the lumen of the duct (fig. 2). In the vestibule, neurofibromas can fix the stapes; this has been pointed out by *Hallpike* [1963] (fig. 4). It is possible that these patients, in a given moment, present with symptoms, which may resemble Ménière's disease as happens in individuals with solitary intralabyrinthine neurinomas [*De Lozier* et al., 1979]. In the facial nerve it is

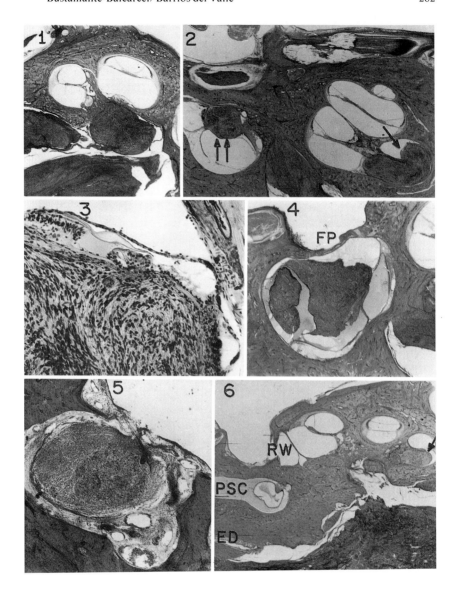

Fig. 1. Three neurofibromas in the internal auditory canal. A tumor arising in the cochlear branch can be appreciated.

Fig. 2. Neurofibromas in the internal auditory canal, in the cochlea (arrow), and in the horizontal semicircular duct (two arrows).

Fig. 3. A neurofibroma is displacing the organ of Corti toward the scala vestibuli.

Fig. 4. Neurofibroma in the vestibule, almost in contact with the foot-plate (FP).

possible to find neurofibromas in the geniculate ganglion as well as in its third portion (fig. 5). These tumors may invade the middle ear. It is also possible to find neurofibromas in its branches; this happened in one of the temporal bones examined by us, where the tumor could be seen in the chorda tympani. Therefore, a lesion of the VIIth nerve in these patients does not necessarily mean that the acoustic nerve is compressed and injured in the angle, a situation that is commonly seen in large neurinomas of the acoustic nerve. It is interesting to mention that although we found histological lesion of the facial nerve, clinically it was not detected. It is possible to find tumors in those nerves that run along the carotid artery in its canal. *Nager* [1969] has demonstrated this and in our series there was a small tumor in one of the temporal bones. A characteristic of VRD is the presence of tumors of several types, particularly if neurofibromas and meningiomas are combined, as well as different types of gliomas [*Wander and Das Gupta*, 1977]. Simultaneous existence of neurofibromas and meningiomas in the temporal bone has been demonstrated by *Nager* [1969] and *Linthicum* [1980]. This author has emphasized the fact that good hearing is present in these patients despite the presence of large tumors [*Linthicum*, 1972]. The study of these tumors by histochemical methods with silver impregnation techniques gives a basic explanation of this phenomenon, demonstrating good preservation of nerve fibers while the tumor was surrounding but not compressing the nerve as is seen in solitary neurinomas of the VIIIth nerve. In our series of temporal bones it has not been possible to identify intact nerve fibers surrounded by tumor. *Nager* [1969] points out that neurofibromas in VRD have greater osteolytic activity than solitary neurofibroma of the VIIIth nerve. In our series we have been able to demonstrate a marked degree of destruction of the internal auditory canal and in 1 case the tumor destroyed the temporal bone in such a way that it was in contact with the endolymphatic duct (fig. 6). In another case the tumor was very close to the carotid canal. Considering the findings in temporal bones from patients with VRD, it is possible to postulate that symptoms in this area can start in many different forms, since it will depend on the initially developing tumor which

---

Fig. 5. Neurofibroma in the third portion of the facial nerve.

Fig. 6. The tumor destroys the temporal bone; it is in contact with the endolymphatic duct.

may begin in the labyrinth, cochlea, internal auditory canal or middle ear. Surgical treatment is more complicated considering that there might be more than one tumor in the internal auditory canal and that their extension in the temporal bone is greater due to their marked osteolytic activity which is very apparent in some cases. This disease is not limited to the internal auditory canal. It may be present in different areas of the temporal bone. To expect good results in terms of preservation of hearing seems to be impossible at the present time.

References

De Lozier, H.L.; Gacek, R.R.; Dana, S.T.: Intralabyrinthic schwannoma. Ann. Otol. Rhinol. Lar. *88:* 187–191 (1979).

Hallpike, C.S.: Stapes fixation by an intra-labyrinthine 'seedling' neurofibroma as a cause of conductive deafness in a case of von Recklinghausen's disease. Acta otolar., suppl. 62–65 (1963).

Linthicum, F.H.: Unusual audiometric and histologic findings in bilateral acoustic neurinomas. Ann. Otol. Rhinol. Lar. *81:* 433–437 (1972).

Linthicum, F.H.: Bilateral acoustic tumors. A diagnostic and surgical challenge. Archs Otolar. *106:* 729–733 (1980).

Nager, C.T.: Acoustic neurinomas. Pathology and differential diagnosis. Archs Otolar. *89:* 252–279 (1969).

Wander, J.V.: Das Gupta, T.: Neurofibromatosis. Curr. Probl. Surg. *14:* 1–18 (1977).

A. Bustamante-Balcarcel, MD, Department of Otolaryngology,
Instituto Nacional de Cardiología 'Ignacio Chávez', México 22, DF (Mexico)

Ataxia

Adv. Oto-Rhino-Laryng., vol. 30, pp. 285–287 (Karger, Basel 1983)

What Distinguishes the Different Kinds of Postural Ataxia in Patients with Cerebellar Diseases[1]

J. Dichgans[a], *H.C. Diener*[a], *K.H. Mauritz*[b]

[a] Department of Neurology at Tübingen, and
[b] Department of Neurology at Freiburg, FRG

Postural imbalance is one of the classical symptoms of cerebellar disease. Methods were developed [*Hufschmidt* et al., 1980; *Dichgans* et al., 1976; *Diener,* 1982; *Diener* et al., 1982] to quantify, analyze, and document postural sway and to differentiate patterns that are typical for the different types of lesions encountered within the cerebellum. Methods entail measurements of the body's center of force (CFP), as projected onto a force-recording platform, the angle versus gravity plus the acceleration of hip and head, the ankle angle, and EMG recordings from the anterior tibial and the triceps surae muscles. Sway path, sway area, mean amplitude of sway, histograms of sway direction and Fourier spectra of sway are computed with the subject standing still or while imposing a postural disturbance by means of unexpectedly tilting or sinusoidally moving the supporting platform.

The results of recordings in patients with degeneration of the spinocerebellar tract, lesions of the cerebellar anterior lobe, the cerebellar hemispheres, and vestibulocerebellar lesions were compared to those of normals.

Patients with afferent *spinocerebellar tract and posterior column lesions* (Friedreich's ataxia) exhibit a low frequency sway with most of its power below 1.1 Hz and a large amplitude. As compared to alcoholics with a cerebellar anterior lobe atrophy, the lateral sway component is rather large in patients with Friedreich's ataxia. Visual stabilization is not as excessive as in some of the alcoholics while standing on stable

[1] Supported by the Deutsche Forschungsgemeinschaft Di 278/1-1.

ground. Sway of head, hip, and CFP may be heavily desynchronized containing quite different dominant frequencies in Friedreich's ataxia while they are synchronized, although opposite in phase for hip and head in patients with anterior lobe atrophy. The most specific finding in patients with Friedreich's ataxia is the massive delay of the stabilizing response of the anterior tibial muscle when the supporting platform is suddenly tilted toe-up (137 ± 32 ms in normals, 296 ± 71 ms in patients). The stabilizing long loop response is also missing in the shortened triceps surae with a tilt toe-down. Vision is very useful to stabilize posture after platform displacement. With their eyes closed, these patients cannot remain upright when the platform is moving. In general patients, need at least two of their three posture-stabilizing sensory channels (proprioception, vision, labyrinth) as soon as the ground becomes unstable. This is also true for patients with a bilateral vestibulectomy [*Diener*, 1982].

Cerebellar anterior lobe (spinocerebellar) lesions are mostly due to alcoholism and malnutrition (atrophie cérébelleuse tardive). This disease almost exclusively involves the Purkinje cells of the medial and intermediate part of the anterior lobe. Most specific for this group is the predominance of anterior-posterior sway–and a conspicuous high frequency component at 2.2.–4.0 Hz [*Silverskiöld*, 1968; *Dichgans* et al., 1976; *Mauritz* et al., 1979, 1981]. The tremor is provoked by eye closure or by suddenly tilting the platform preferably toe-up. Its frequency was reported to decrease with progression of the disease [*Mauritz*, 1979]. Long latency reflexes, although normal in latency are increased in amplitude and duration [*Diener*, 1982]. This and the obvious increase in amplitude of intersegmental movements (head/trunk and trunk/hip, etc.) may explain the tremor. Despite the tremor and a considerable increase in sway path and mean amplitude of sway, these patients are generally able to remain upright, even with rapid tilts of the supporting platform. Visual stabilization is most pronounced.

Lesions of the *vestibulocerebellum* in patients with medulloblastomas cause an omnidirectional sway of excessive amplitude predominantly containing frequencies around and below 1 Hz. Intersegmental movements are diminished [*Mauritz* et al., 1979]. Patients frequently fall, even with their eyes open, also when sitting and without an external disturbance. It is assumed that the vestibular graviceptive set value for orientation versus gravity is inaccessible for postural stabilization causing drifts and poor control activity, whereas with anterior lobe lesions the correct position seems to be known to the system, but the control loops oscillate because of increased gain and poor control of the duration of the response.

Tumors of the *cerebellar hemispheres* – the pontocerebellum or neocerebellum – which controls volitional movements of the upper more than the lower extremities cause no or only small disturbances of stance and show no characteristic abnormality with the methods used. Sway is omnidirectional, its frequency spectrum is normal. *Mauritz* [1979] demonstrated abnormal deviations in patients who attempted to follow with their CFP displayed on an oscilloscope, a sinusoidally moving target displayed on the same scope. This deficiency fits the assumption that the cerebellar hemispheres subserve the fine spatiotemporal organization of pursuit.

References

Dichgans, J.; Mauritz, K.H.; Allum, J.H.J.; Brandt, T.: Postural sway in normals and atactic patients: Analysis of the stabilizing and destabilizing effects of vision. Agressologie *17:* 15–24 (1976).

Diener, H.C.; Dichgans, J.; Bruzek, W.; Selinka, H.: Stabilization of human posture during induced oscillations of the body. Exp. Brain Res. *45:* 126–132 (1982).

Diener, H.C.: Zur Physiologie und Pathophysiologie des aufrechten Stehens beim Menschen unter statischen und dynamischen Bedingungen; med. Habilitationsschr., Universität Tübingen (1982).

Hufschmidt, A.; Dichgans, J.; Mauritz, K.H.; Hufschmidt, M.: Some methods and parameters of body sway quantification and their neurological applications. Arch. Psychiat. NervKrank. *228:* 135–150 (1980).

Mauritz, K.H.: Standataxie bei Kleinhirnläsionen; med. Habschr., Universität Freiburg (1979).

Mauritz, K.H.; Dichgans, J.; Hufschmidt, A.: Quantitative analysis of stance in late cortical cerebellar atrophy of the anterior lobe and other forms of cerebellar ataxia. Brain *102:* 461–482 (1979).

Mauritz, K.H.; Schmidt, C.; Dichgans, J.: Delayed and enhanced long latency reflexes as the possible cause of postural tremor in late cerebellar atrophy. Brain *104:* 97–116 (1981).

Silverskiöld, B.P.: Romberg's test in the cerebellar syndrome occuring in chronic alcoholism. Acta neurol. scand. *45:* 292–302 (1968).

J. Dichgans, MD, Department of Neurology at Tübingen,
Liebermeisterstrasse 18-20, D-7400 Tübingen (FRG)

Adv. Oto-Rhino-Laryng., vol. 30, pp. 288–290 (Karger, Basel 1983)

Functional Plasticity of Spinal and Supraspinal Reflexes in Maintaining Upright Stance[1]

H.C. Diener, J. Dichgans, F. Bootz

Neurologische Universitätsklinik, Tübingen, FRG

The maintenance of upright posture requires the integration of afferent information from proprioceptors, vision, and the vestibular system and the coordination of early spinal and supraspinal reflexes with subcortical postural and late voluntary control movements. Segmental stretch reflexes in the leg muscles alone are not able to compensate for an external disturbance applied to the body [*Allum and Büdingen,* 1979]. Electromyographic (EMG) responses of longer latencies are required in addition to restore the upright position. Inertial forces and the viscoelastic properties of leg muscles, tendons, and ligaments also contribute. Our experiments were undertaken in order to investigate the timing of short, medium, and long latency EMG responses and to explore their functional plasticity under different demands [*Diener,* 1982].

Subjects stood with open eyes on a force-measuring platform which could be rotated in pitch around an axis collinear with the subjects ankle joint. The coordinates of the center of foot pressure were determined by means of strain gauges at the four corners of the platform. Head and hip displacements were recorded with goniometers. The rate of ramp platform movements toe-up or toe-down varied between 5 and 100°/s, the amplitude between 1 and 8°. Surface EMGs of the anterior tibial and the triceps surae muscles were full wave rectified. EMG latencies were measured from the beginning of the displacement of the platform. Eight runs were recorded from each stimulus condition.

[1] Supported by the Deutsche Forschungsgemeinschaft Di 278/1-1.

A sudden upward displacement of the platform causes a segmental stretch reflex in the triceps surae with a mean latency of 50–60 ms. With increasing rate ($> 40°/s$) and amplitude ($> 4°$) of platform movement a medium latency response with a latency of 115–125 ms can be observed in the stretched muscle. Both reflexes are functionally destabilizing, since they support the induced passive backward displacement of the body. Compensation of body displacement in this situation is corroborated by a late EMG response in the anterior tibial muscle with a mean latency of 130–140 ms. Tilting the platform down never evoked a segmental stretch reflex, but a medium latency response in the anterior tibial muscle with a mean latency of 110–120 ms. The late, functionally stabilizing response in the triceps surae can be observed after 150–170 ms. The latencies of the EMG responses were independent of the amplitude of the platform movement (at $80°/s$). The latencies reached a minimum at $40°/s$ (with an amplitude of $4°$).

In accordance with *Gottlieb and Agarwal* [1979] the integrals of rectified EMG from both muscles were linearly related to both, the amplitude ($2–6°$) and the rate ($20–100°/s$) of the platform displacement. The slope of these functions (gain) was invariably higher for the late, functionally stabilizing response in the antagonist than in the stretched muscles (short and medium latency responses).

The functional plasticity of segmental and suprasegmental reflexes was investigated by changing the initial body position prior to a ramp platform movement. 10 subjects set their body position by leaning backwards or forwards while standing. The set was visually controlled by matching the CFP with a target on an oscilloscope indicating a displacement of 5 cm. While leaning backwards a tilt of the platform toe-up leads to suppression of the stretched evoked segmental response in the triceps surae, whereas the medium latency response, although dangerous, in this position is preserved. The latency of the functionally stabilizing response in the anterior tibial muscle is shortened. This causes a cocontraction of the two antagonistic muscles thereby neutralizing the destabilizing action of the triceps surae. In a similar manner, after leaning forwards and tilting the platform toe-down a cocontraction of both muscles can be observed, but here the triceps surae muscle has to compensate the destabilizing effects of the activity of the anterior tibial muscle.

These latency changes show the plasticity of segmental and suprasegmental reflex mechanisms according to functional requirements.

We agree with *Nashner and Cordo* [1982], who stress that 'postural' responses are largely independent of voluntary control. Voluntary movements are unlikely, since latencies were not shortened by training and intraindividual standard deviations were too small. Subjects were not informed about the time, direction, rate, and amplitude of the platform displacement. The late EMG responses could, however, still be 'triggered' by a setting of the postural stabilizing system preparing fixed responses for leaning forwards and backwards respectively. The functionally useful response then would be triggered by the stretch and/or sudden shortening of the two muscles under each condition.

References

Allum, J.H.J.; Büdingen, H.J.: Coupled stretch reflexes in ankle muscles: an evaluation of the contribution of active muscle mechanisms to human posture stability. Prog. Brain Res. *50:* 185–195 (1979).

Diener, H.C.: Zur Physiologie und Pathophysiologie des aufrechten Stehens beim Menschen unter statischen und dynamischen Bedingungen; Med. Habschr., Universität Tübingen (1982).

Gottlieb, G.L.; Agarwal, G.C.: Response to sudden torques about ankle in man. J. Neurophysiol. *42:* 91–106 (1979).

Nashner, L.M.; Cordo, P.J.: Relation of automatic postural responses and reaction-time voluntary movements of human leg muscles. Exp. Brain Res. *43:* 395–405 (1982).

H.C. Diener, MD, Department of Neurology at Tübingen,
Liebermeisterstrasse 18–20, D-7400 Tübingen (FRG)

Adv. Oto-Rhino-Laryng., vol. 30, pp. 291–297 (Karger, Basel 1983)

Ataxia and Oscillopsia in Downbeat-Nystagmus Vertigo Syndrome[1]

W. Büchele, T. Brandt, D. Degner

Neurological Clinic with Clinical Neurophysiology, Alfried Krupp Hospital, Essen, FRG

Downbeat nystagmus in the primary or lateral position of gaze is a well-defined clinical sign, almost specific for a structural lesion of the paramedian craniocervical junction [*Cogan,* 1968]. The suggested mechanism of a downward pursuit defect [*Zee* et al., 1974] was recently questioned by *Baloh and Spooner* [1981] who assume 'that downbeat nystagmus is a type of central vestibular nystagmus resulting from an imbalance in the central vertical vestibulo-ocular pathways, either by a lesion in the floor of the 4th ventricle between the vestibular nuclei (which interrupts the tonic excitatory activity to the inferior recti), or by a lesion of the flocculus (which leads to an increase in tonic excitatory activity to the superior recti due to a disinhibition)'.

The clinical oculomotor abnormalities have been frequently described: downbeat nystagmus is present in darkness as well as with fixation; amplitude and slow phase velocity increase on lateral gaze or with head extension; downward 'pursuit' is saccadic. The patients, however, complain of a distressing illusory oscillation of the visual scene (oscillopsia) and postural imbalance, both obligate, but hitherto poorly studied, symptoms of the syndrome. The retinal slip in downbeat nystagmus is misinterpreted as motion of the visual scene, because the involuntary ocular movements which override fixation are not associated with an appropriate efference-copy signal (fig 1). The distur-

[1] Supported by the Deutsche Forschungsgemeinschaft, Br 639/5.

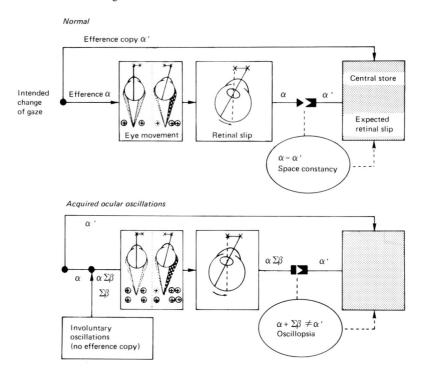

Fig. 1. Schematic pathomechanism of ocular vertigo and oscillopsia with acquired ocular oscillations such as downbeat nystagmus. In normals the voluntary impulse to perform a change of gaze releases the efference α to the eye muscles as well as an appropriate efference copy signal α' to a central store. This store contains the memory for the expected retinal slip due to the particular intended eye movement as calibrated prior to the disease onset. Space constancy is maintained if the comparison of the actual with the expected retinal slip is $\alpha = \alpha'$. Involuntary ocular oscillations which override fixation are not associated with an appropriate efference copy signal and therefore cause oscillopsia, an illusory oscillation of the visual scene. Oscillopsia causes an impairment of visual stabilization of balance since retinal image motion is a major cue for body stabilization.

bance of spatial localization as well as the oscillopsia are not restricted to the fovea but involve the entire visual field and therefore affect the two modes of visual processing 'focal' and 'ambient', respectively. The ambient mode relies on afferent information from the peripheral field and subserves postural balance. Consequently, oscillopsia should cause an impairment of postural balance since retinal image motion is a major cue for body stabilization.

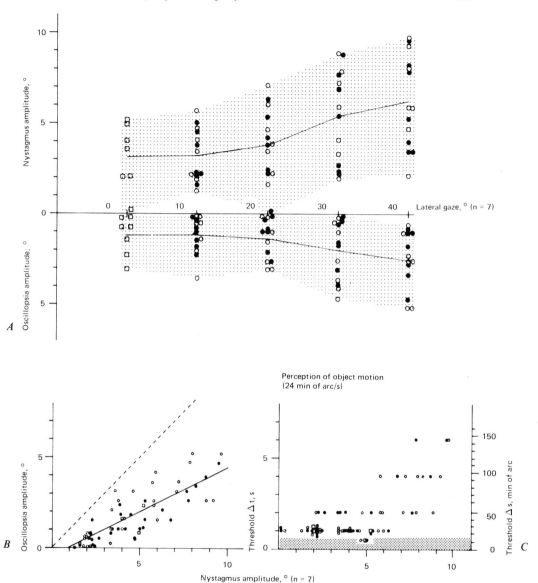

Fig. 2. A Mean amplitudes of downbeat nystagmus and concurrent subjective oscillopsia as a function of lateral gaze in 7 patients (O = gaze right; ● = gaze left). *B* Oscillopsia as a function of nystagmus amplitude. *C* Thresholds for the perception of object
motion as a function of nystagmus amplitude. Shaded area represents mean thresholds
of normals. Oscillopsia is always smaller than the net retinal slip, increases with increasing nystagmus amplitude with a mean ratio of 0.37.

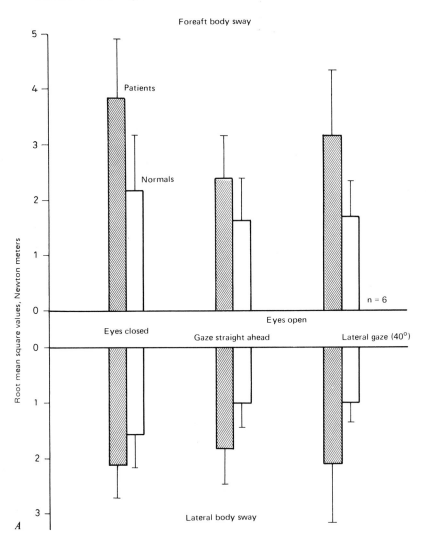

Fig. 3. A Posturography of the fore-aft and lateral body sway (root mean square values, Newton meters) in 6 patients with downbeat nystagmus (shaded columns) as compared to a control group of normals (open columns). Vestibulocerebellar ataxia is manifest in the abnormal fore-aft body sway with the eyes closed. Impaired visual control of postural balance is particularly apparent on lateral gaze which activates the nystagmus intensity. *B* Simultaneous recordings of vertical eye movements and body sway in a patient at free stance with the head upright and fixating a target either straight ahead or 40° lateral.

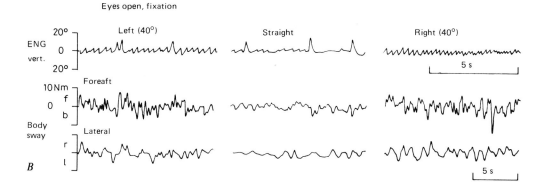

Simultaneous psychophysical, electronystagmographic and postu-
rographic measurements have been performed in 7 patients in order to
elucidate the relationship between retinal image slip, oscillopsia and
postural imbalance:

Oscillopsia is a permanent symptom but the illusory motion is
smaller than that which would be expected from the amplitude of the
nystagmus; oscillopsia is dependent on the direction of gaze as is nystagmus amplitude; it increases with increasing amplitude (fig. 2) with a
mean ratio between the two of 0.37; the individual ratio is relatively
consistent for each patient with an interindividual range from 0.13 to
0.61.

Thresholds for egocentric detection of object motion are significantly
raised in patients with downbeat nystagmus as compared to healthy
subjects; thresholds increase with increasing nystagmus amplitude
(fig. 2c). Thus, there is a partial suppression of visual motion perception for both the retinal slip due to the involuntary eye movements as
well as for single objects moving within the visual scene. This 'adaptive
suppression' is beneficial to the organism to the extent that this allevi-
ates the distressing oscillopsia with the disadvantageous side effect,
however, of increased thresholds for egocentric perception of object
motion in general.

Postural imbalance is particularly apparent for the fore-aft body
sway with a tendency to fall backward. This fore-aft postural instability can be interpreted as a direction specific vestibulocerebellar imbalance since it can be observed with the eyes closed (fig. 3). We believe

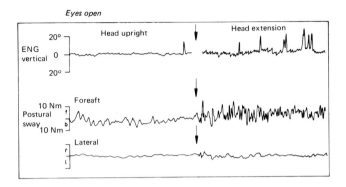

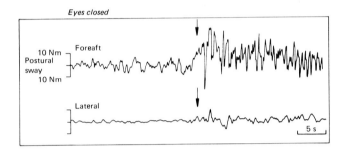

Fig. 4. Postural instability in downbeat nystagmus is not simply due to ocular vertigo because there is a marked fore-aft ataxia even with the eyes closed, which becomes particularly apparent with head extension (bottom). Moreover, with the eyes open, visual control of balance is discernable but somewhat diminished by concurrent activation of the downbeat nystagmus consequent to head extension as can be seen by simultaneous recordings of vertical ENG and body sway.

that the objective measurable backward tilt represents a vestibulospinal compensation in the direction opposite to the perceived lesional forward vertigo which corresponds to the downbeat nystagmus. With the eyes open a measurable visual stabilization of body sway is preserved but does not sufficiently compensate for the vestibular ataxia (fig. 4). In downbeat nystagmus the patient's pathological postural sway, with the eyes open, is dependent on the direction of gaze; it increases with increasing nystagmus amplitude; and pathophysiologically it is secondary to a combination of both a vestibulocerebellar ataxia and a reduced visual stabilization owing to the nystagmus.

References

Baloh, R.W.; Spooner, J.W.: Downbeat nystagmus: a type of central vestibular nystagmus. Neurology *31:* 304–310 (1981).

Cogan, D.G.: Down-beat nystagmus. Archs. Ophthal. *80:* 757–768 (1968).

Zee, D.S.; Friendlich, A.R.; Robinson, D.A.: The mechanism of downbeat nystagmus. Archs Neurol. *30:* 227–237 (1974).

Dr. W. Büchele, Neurological Clinic with Clinical Neurophysiology, Alfried Krupp Hospital, Alfried Krupp Strasse 21, D-4300 Essen (FRG)

Adv. Oto-Rhino-Laryng., vol. 30, pp. 298–301 (Karger, Basel 1983)

Frequency Response of Medullary Reticulospinal Neurons to Natural Stimulation of Labyrinth Receptors

D. Manzoni, O. Pompeiano, G. Stampacchia, U.C. Srivastava

Istituto di Fisiologia Umana, Cattedra I, Università di Pisa, Italia

Previous experiments have shown that during static or dynamic tilts the labyrinth reflexes may act asymmetrically on limb extensors in that side-down rotation of the head after eliminating neck reflexes [4] or side-down tilt of the whole animal [8] produces contraction, whereas side-up rotation results in extensor relaxation. These responses have been attributed, at least in part, to activity of neurons located in the lateral vestibular nucleus (LVN) of Deiters, whose descending vestibulospinal projection exerts a prominent excitatory influence on ipsilateral extensor motoneurons [5]; in fact most of the LVN units which respond preferentially to the direction of orientation of the angular stimulus, i.e. to the extreme animal position, exhibit an increase in firing rate during side-down tilt and a decrease in firing rate during side-up tilt (α-response), while a smaller proportion of units show the opposite response pattern (β-response) [1, 2]. Similar results were also reported for the responses to tilt of the primary afferents originating from otolith receptors in the utriculus, thus in apparent agreement with the morphological polarization of the receptors [cf. 1 for references].

The main aim of the present experiments was to find out whether presumably inhibitory reticulospinal neurons collaborate with excitatory vestibulospinal neurons to the motoneuronal output controlling limb extensors during the labyrinth reflexes.

The activity of 168 medullary reticular neurons histologically located in the ventromedial aspect of the medullary reticular formation, i.e. in that area which upon stimulation produced inhibitory postsynaptic potentials in ipsilateral hindlimb motoneurons [3], was recorded extracellularly in decerebrate cats, which were paralyzed and artificially ventilated. Among these units 93 were reticulospinal neurons activated antidromically by spinal cord stimulation at T_{12}–L_1 (lRS neurons), while the remaining 75 neurons were not activated by spinal cord stimulation (RF neurons). Each unit was tested during sinusoidal rotation about the longitudinal axis of the whole animal at the standard parameters (0.026 Hz, ±10° peak amplitude) leading to stimulation of labyrinth receptors [1]. The unit activity was processed on-line with a computer system equipped with a fast Fourier analyzer. A spectral analysis of the angular input (table rotation) and of the output (unit activity) was performed and the gain (imp/s/°), sensitivity (% increase of the base discharge frequency per degree of peak displacement), and phase angle of the first harmonic of response were evaluated, the latter being expressed in arc degrees between the peak of the fundamental component of response and the peak of the side-down displacement of the animal, indicated by 0°.

113 out of 168 (67.3%) units tested showed periodic modulation of firing rate in response to the sinusoidal input. 71 (62.8%) units were excited by side-up tilt (β-response) and 24 (21.2%) units by side-down tilt (α-response), both populations of neurons firing with an average phase lead of the first harmonic of response of +25.3±28.3°, SD, with respect to the extreme animal displacements. In the remaining 18 (15.9%) units the response peak was related to the velocity vector of movement (table I).

The average gain of responses corresponded to 0.32±0.36, SD, imp/s/°, whereas the average sensitivity expressed in % increase of the base discharge frequency per degree corresponded to 3.80±4.08, SD (base frequency, 11.4±9.6, SD, imp/s, n = 113). Similar results were obtained from the two populations of lRS and RF neurons (table I).

Responses of reticulospinal neurons to tilt were still observed at the peak amplitude of 1°. Gain and sensitivity of responses did not greatly change by increasing amplitude of tilt from 1 to 20° (at 0.026 Hz), indicating that the system was linear with respect to the amplitudes used. By varying frequency of tilt from 0.008 to 0.32 Hz (at ±10°) two populations of neurons were observed. In the first the sensitivity and phase angle of responses remained unmodified against changes in frequency, due to stimulation of macular receptors. In the second population the sensitivity of responses increased by increasing

Table I. Comparison of the responses of medullary reticular formation neurons to sinusoidal roll tilt of the animal (at 0.026 Hz, ± 10°) with those of LVN neurons recorded in previous experiments [1]

	IRS	RF	Total	rvLVN	dcLVN	Total
Number of units	93	75	168	34	68	102
Responsive (R) units	64 (68.8)	49 (65.3)	113 (67.3)	31 (91.2)	46 (67.6)	77 (75.5)
Nonresponsive units	29 (31.2)	26 (34.7)	55 (32.7)	3 (8.8)	22 (32.4)	25 (24.5)
Base discharge frequency of R units (mean ± SD)	10.8 ± 9.1	12.2 ± 10.2	11.4 ± 9.6	45.6 ± 68.2	43.2 ± 48.9	43.6 ± 54.4
Gain of R units (mean ± SD)	0.30 ± 0.32	0.36 ± 0.40	0.32 ± 0.36	0.58 ± 0.58	0.41 ± 0.35	0.48 ± 0.44
Sensitivity of R units (mean ± SD)	3.66 ± 3.64	4.12 ± 4.63	3.80 ± 4.08	3.62 ± 3.40	1.81 ± 1.83	2.54 ± 2.71
	n = 61	n = 46	n = 107	n = 31	n = 46	n = 77
Patterns of response						
α – Response	21 (32.8)	3 (6.1)	24 (21.2)	13 (41.9)	19 (41.3)	32 (41.6)
β – Response	37 (57.8)	34 (69.4)	71 (62.8)	8 (25.8)	7 (15.2)	15 (19.5)
Intermediate responses	6 (9.4)	12 (24.5)	18 (15.9)	10 (32.3)	20 (43.5)	30 (38.9)

IRS and RF = Medullary reticular neurons activated or non-activated antidromically by spinal cord stimulation at T_{12}–L_1; rvLVN, dcLVN = neurons located in the rostroventral and the dorsocaudal part of Deiters nucleus projecting to cervical and lumbosacral segments of the spinal cord, respectively; base discharge frequency, in imp/s; gain of first harmonic of responses, in imp/s/°; sensitivity, in imp/s/° in % of the base frequency. Numbers in parentheses are percentages.

frequency of rotation above 0.051 Hz, while the phase lead of responses became related to angular velocity. These effects were attributed to costimulation of vertical canal receptors.

In summary, it appears that the predominant response pattern of the medullary reticulospinal neurons to roll tilt (β-response) is just opposite to that of the lateral vestibulospinal neurons (α-response). This finding is attributed to the fact that the macular input of one side can be transmitted to the contralateral medullary reticular formation by utilizing the lateral vestibulospinal tract acting on neurons of the crossed spinoreticular pathway [cf. 6, 7 for references].

The reciprocal pattern of responses of the reticulospinal and the vestibulospinal neurons to the same parameters of tilt suggests that the former neurons exert an inhibitory influence on ipsilateral hindlimb extensor motoneurons, in contrast to the latter neurons which are excitatory in function. If this hypothesis is correct, then the motoneurons innervating the ipsilateral limb extensors would be not only excited during side-down tilt by an increased discharge of vestibulospinal neurons but also disinhibited by a reduced discharge of reticulospinal neurons; the opposite would occur during side-up tilt. It is tempting to suggest that the reticulospinal system intervenes in the gain regulation of response of extensor motoneurons to stimulation of the vestibulospinal reflex arc.

Reticulospinal neurons located in the inhibitory area of the medullary reticular formation may thus contribute with excitatory vestibulospinal neurons to the postural adjustments of hindlimb muscles during the labyrinth reflexes.

References

1   Boyle, R.; Pompeiano, O.: Reciprocal responses to sinusoidal tilt of neurons in Deiters' nucleus and their dynamic characteristics. Archs ital. Biol. *118*: 1–32 (1982).
2   Boyle, R.; Pompeiano, O.: Convergence and interaction of neck and macular vestibular inputs on vestibulospinal neurons. J. Neurophysiol. *45*: 852–868 (1981).
3   Jankowska, E.; Lund, S.; Lundberg, A.; Pompeiano, O.: Inhibitory effects evoked through ventral reticulospinal pathways. Archs ital. Biol. *106*: 124–140 (1968).
4   Lindsay, K.W.; Roberts, T.D.M.; Rosenberg, J.R.: Asymmetric tonic labyrinth reflexes and their interaction with neck reflexes in the decerebrate cat. J. Physiol., Lond. *261*: 583–601 (1976).
5   Lund, S.; Pompeiano, O.: Monosynaptic excitation of alpha motoneurones from supraspinal structures in the cat. Acta physiol. scand. *73*: 1–21 (1968).
6   Pompeiano, O.: Macular input to neurons of the spinoreticulocerebellar pathway. Brain Res. *95*: 351–368 (1975).
7   Pompeiano, O.: Neck and macular labyrinthine influences on the cervical spinoreticulocerebellar pathway. Prog. Brain Res. *50*: 501–514 (1979).
8   Schor, R.H.; Miller, A.D.: Vestibular reflexes in neck and forelimb muscles evoked by roll tilt. J. Neurophysiol. *46*: 167–178 (1981).

O. Pompeiano MD, Istituto di Fisiologia Umana, Cattedra I, Università di Pisa, Via S. Zeno 31, I-56100 Pisa (Italy)

Adv. Oto-Rhino-Laryng., vol. 30, pp. 302–305 (Karger, Basel 1983)

Frequency Response of Medullary Reticulospinal Neurons to Sinusoidal Rotation of the Neck

U.C. Srivastava, D. Manzoni, O. Pompeiano, G. Stampacchia

Istituto di Fisiologia Umana, Cattedra I, Università di Pisa, Italia

The tonic cervical reflexes originating from neck joint and/or muscle spindle receptors may act asymmetrically on limb extensors; in fact, side-down rotation of the cervical axis vertebra about the longitudinal axis of the animal produces relaxation of the extensors – associated with active flexion – of the ipsilateral limbs, whereas side-up rotation of the neck results in the opposite posture [1, 2]. These responses depend, at least in part, upon the activity of vestibulospinal neurons originating from the lateral vestibular nucleus (LVN) of Deiters, which are excitatory on ipsilateral extensor motoneurons [8]. In fact most of the LVN units, which respond to the direction of neck orientation and not to the angular velocity of neck rotation, exhibit a decrease in firing rate during side-down neck rotation, and an increase in firing rate during side-up neck rotation; on the other hand a proportion of units show the opposite response pattern [1] (table I). Neurons exhibiting the predominant response pattern were found particularly within the dorso-Caudal part of Deiters nucleus (dcLVN), which projects to the lumbosacral segments of the spinal cord [10].

The main aim of the present experiments was to find out whether presumably inhibitory reticulospinal neurons collaborate with excitatory vestibulospinal neurons to the motoneuronal output controlling the hindlimb musculature during the neck reflexes.

The activity of 132 medullary reticular neurons, histologically located in the ventromedial aspect of the medullary reticular formation, i.e. in that area which upon stimulation inhibited ipsilateral hindlimb motoneurons [7], was recorded extracellularly in decerebrate cats. The animals were paralyzed and artificially ventilated. Among these units

Table I. Comparison of the responses of medullary reticular formation neurons to sinusoidal neck rotation (at 0.026 Hz, ± 5° or 10°) with those of LVN neurons recorded in previous experiments [1]

	IRS	RF	Total	rvLVN	dcLVN	Total
Number of units	85	47	132	46	74	120
Responsive (R) units	66 (77.6)	31 (66.0)	97 (73.5)	34 (73.9)	36 (48.6)	70 (58.3)
Nonresponsive units	19 (22.4)	16 (34.0)	35 (26.5)	12 (26.1)	38 (51.4)	50 (41.7)
Base discharge frequency						
of R units (mean ± SD)	10.4 ± 8.2	13.1 ± 11.1	11.3 ± 9.2	38.6 ± 64.8	41.4 ± 38.5	40.7 ± 48.9
Gain of R units (mean ± SD)	0.52 ± 0.45	0.32 ± 0.27	0.45 ± 0.41	0.45 ± 0.58	0.75 ± 0.82	0.60 ± 0.70
Sensitivity of R units	5.70 ± 4.73	3.35 ± 3.42	4.90 ± 4.47	3.81 ± 4.54	2.73 ± 2.84	3.26 ± 3.77
(mean ± SD)	n = 61	n = 30	n = 91	n = 34	n = 36	n = 70
Patterns of response						
Side-down neck rotation	47 (71.2)	23 (74.2)	70 (72.2)	14 (41.2)	9 (25.0)	23 (32.8)
Side-up neck rotation	13 (19.7)	6 (19.35)	19 (19.6)	9 (26.5)	18 (50.0)	27 (38.6)
Intermediate responses	6 (9.1)	2 (6.45)	8 (8.2)	11 (32.3)	9 (25.0)	20 (28.6)

IRS and RF = medullary reticular neurons activated or non-activated antidromically by spinal cord stimulation at T_{12}-L_1; rvLVN and dcLVN, neurons located in the rostroventral and the dorsocaudal part of Deiters nucleus projecting to cervical and lumbosacral segments of the spinal cord, respectively; base discharge frequency, in imp/s; gain of first harmonic of responses, in imp/s/°; sensitivity, in imp/s/° in % of the base frequency. Numbers in parentheses are percentages.

85 were reticulospinal neurons activated antidromically by electrical stimulation of the spinal cord at T_{12}-L_1 (IRS neurons), while the remaining 47 neurons could not be antidromically identified (RF neurons). Each unit was tested during sinusoidal displacement of the neck at the standard parameters (0.026 Hz, ±10° peak amplitude). This was achieved by rotation of the body about the longitudinal axis of the animal, while maintaining the head stationary. A Fourier analysis of the responses was performed. In particular the gain (imp/s/°), sensitivity (% increase of the base discharge frequency per degree of peak displacement) and phase angle of the first harmonic of response were evaluated, the latter being expressed in arc degrees between the peak of the fundamental component of response and the peak of the side-down displacement of the neck indicated by 0°.

97 out of 132 (73.5%) units tested responded with a periodic modulation of firing rate to sinusoidal neck rotation. Most of these units were excited during side-down and inhibited during side-up rotation

(70 units, 72.2%), while a smaller group (19 units, 19.6%) showed the opposite behavior. These populations of units showed an average phase lead of the first harmonic of response of $+41.0 \pm 19.2°$, SD, with respect to the extreme neck displacements. The remaining 8 (8.2%) units showed a response peak which was related to velocity of neck rotation (table I).

The average gain of responses corresponded to 0.45 ± 0.41, SD, imp/s/°, whereas the average sensitivity expressed in % increase of the base discharge frequency per degree corresponded to 4.90 ± 4.47, SD (base frequency, 11.3 ± 9.2, SD, imp/s, n = 97). There were no great differences in the average gain and sensitivity of the responses between the two populations of antidromically activated and nonantidromic neurons (table I).

Responses of reticulospinal neurons were still observed at the peak amplitude of neck rotation of 0.25°. Gain and sensitivity of responses decreased by increasing the peak amplitude of neck rotation from 0.5 to 10° (at 0.026 Hz). Changes in frequency of stimulation from 0.008 to 0.051 Hz (at $\pm 10°$) did not modify the gain and the phase angle of the unit responses, indicating that these responses depended on changes in neck position. However, the gain and the phase lead of responses increased by increasing frequency of rotation from 0.051 to 0.32 Hz.

In summary, the present experiments have shown that the predominant response pattern to neck rotation of the medullary reticulospinal neurons projecting to the lumbosacral segments of the spinal cord (excitation during side-down neck displacement), is just opposite to that of the lateral vestibulospinal neurons, which also project to the same segments of the spinal cord. This finding is attributed to the fact that the neck input of one side exerts a direct excitatory influence on the ipsilateral medullary reticular formation, probably by utilizing neurons of the uncrossed spinoreticular pathway. On the other hand an uncrossed spinoreticulo cerebellar pathway would transmit the neck input of one side to Purkinje cells of the ipsilateral cerebellar vermis [3–6], thus inhibiting the corresponding LVN neurons [cf. 9 for references].

The observation that the responses of reticulospinal and vestibulospinal neurons to sinusoidal neck rotation at the standard parameters are about 180° out of phase with respect to each other strongly supports the view that the reticulospinal neurons are inhibitory on ipsilateral hindlimb extensor motoneurons, in contrast to the vestibulospinal neurons which are excitatory in function. It is suggested therefore that during side-down neck rotation the motoneurons innervating the ipsilateral hindlimb extensors are disfacilitated by the reduced discharge of vestibulospinal neurons, but are inhibited by the increased discharge of reticulospinal neurons; the opposite would occur during side-up neck displacement. It is not unlikely to believe that the reticulospinal system intervenes in the regulation of the gain of response of extensor motoneurons to.

Reticulospinal neurons located in the inhibitory area of the medullary reticular formation may thus contribute with excitatory vestibulospinal neurons to the postural changes of the hindlimb musculature during the neck reflexes.

References

1   Boyle, R.; Pompeiano, O.: Responses of vestibulospinal neurons to sinusoidal rotation of the neck. J. Neurophysiol. *44:* 633–649 (1980).

2   Boyle, R.; Pompeiano, O.: Convergence and interaction of neck and macular vestibular inputs on vestibulospinal neurons. J. Neuropyhsiol. *45:* 852–868 (1981).

3   Denoth, F.; Magherini, P.C.; Pompeiano, O.; Stanojević, M.: Responses of Purkinje cells of the cerebellar vermis to sinusoidal rotation of neck. J. Neurophysiol. *43:* 46–59 (1979).

4   Denoth, F.; Magherini, P.C.; Pompeiano, O.; Stanojević, M.: Responses of Purkinje cells of the cerebellar vermis to neck and macular vestibular inputs. Pflügers Arch. *381:* 87–98 (1979).

5   Kubin, L.; Magherini, P.C.; Manzoni, D.; Pompeiano, O.: Responses of lateral reticular neurons to sinusoidal rotation of neck in the decerebrate cat. J. Neuroscience *6:* 1277–1290 (1981).

6   Kubin, L.; Manzoni, D.; Pompeiano, O.: Responses of lateral reticular neurons to convergent neck and macular vestibular inputs. J. Neurophysiol. *46:* 48–64 (1981).

7   Jankowska, E.; Lund, S.; Lundberg, A.; Pompeiano, O.: Inhibitory effects evoked through ventral reticulospinal pathways. Archs ital. Biol. *106:* 124–140 (1968).

8   Lund, S.; Pompeiano, O.: Monosynaptic excitation of alpha motoneurones from supraspinal structures in the cat. Acta physiol. scand. *73:* 1–21 (1968).

9   Pompeiano, O.: Neck and macular labyrinthine influences on the cervical spino-reticulocerebellar pathway. Prog. Brain Res. *50:* 501–514 (1979).

10 Pompeiano, O.; Brodal, A.: The origin of vestibulospinal fibers in the cat. An experimental-anatomical study, with comments on the descending medial longitudinal fasciculus. Archs ital. Biol. *95:* 166–195 (1957).

O. Pompeiano, MD, Istituto di Fisiologia Umana, Cattedra I, Università di Pisa, Via S. Zeno 31, I-56100 Pisa (Italy)

Adv. Oto-Rhino-Laryng., vol. 30, pp. 306–310 (Karger, Basel 1983)

Evaluation of Ataxia with Square Drawing Test Discussing Macrographism

Hideki Ohyama[a], *Shoichi Honjo*[b], *Toru Sekitani*[a],
Keiko Nishikawa[a], Yoshihiko Okinaka[a], *Takaaki Matsuo*[a]

[a] Department of Otolaryngology, Yamaguchi University School of Medicine, Ube City; [b] Hikari City Hospital, Hikari, Japan

Among the various examinations for vestibulospinal reaction, such as the past pointing test, Abweich reaction and the arm tonus reaction, there is the so-called writing test. In Japan there are two prevailing procedures of the writing tests: Fukuda's blindfolded vertical writing test established in 1943 and Sekitani's Square Drawing Test (SDT) established in 1973. These writing tests cannot only prove minute deviations in the fine movement of the finger, hand and arm, through the finger, elbow and shoulder joints and related neuromuscular system, but can also record purely objective findings of every subject.

Method

Five parameters of the SDT have been isolated. The right-hand set containing four squares prescribed in 40×40 mm were drawn with eyes open. The center line also containing four squares, i.e. from R-1 to R-4, were drawn with eyes closed. We can derive five parameters from these four squares; R-1 to R-4 (fig. 1). We have now designed a test series of alcohol affection in order to analyze and investigate the variations of the central nervous system when influenced by the consumption of alcohol. 30 subjects, who were all in their twenties and normal otoneurologically, were divided into two groups, viz. a 'drinking group' consisting of 17 subjects, and a 'nondrinking group' of 13 subjects. Examinations included a spontaneous nystagmus and an eye tracking test involving the vestibuloocular system, as well as a writing test and stabilography involving the vestibulospinal system. These examinations were loaded every 15 min for the following 2 h, and along with a measurement of alcohol concentration in the breath, were done at a set period in the alcohol drinking group (fig. 2).

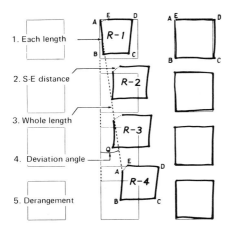

Fig. 1. Parameters of SDT.

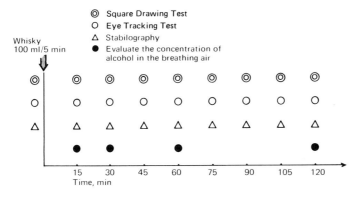

Fig. 2. Test protocol for the influence of alcohol on the vestibular functions, especially in this paper on the performance of SDT.

Results and Comment

The highest concentration of alcohol in the breath was detected at 15 min after drinking and then the titer decreased gradually for the following 2 h. First, the actual SDT data of an interesting case is presented. The left-hand set containing four squares was drawn with eyes open before drinking. A mild shortening but no notable deviation or derangement of the drawn line is seen. The next set to the right (in this

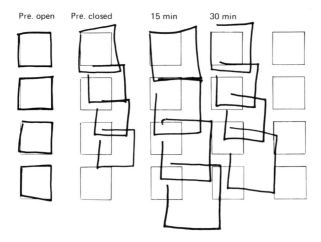

Fig. 3. Actual record of SDT in 1 case who drank 100 ml of whisky (43% alcohol concentration) within 5 min. Note dominant macrographism of SDT after drinking, which appeared at the 15-min period and then remained.

specially arranged figure) was drawn with eyes closed before drinking. These findings are within the normal range.

The whole length is also thought to be within normal limits. The third and fourth sets were drawn 15 and 30 min after drinking with eyes closed. Dominant macrographism, elongation of the whole length and also separation are noted. At the 30-min period, elongation became decreased, though dominant macrographism still remained (fig. 3). There are four sets of squares, from left to right, drawn at 60, 75, 90 and 105 min after drinking, respectively. Data from 60 to 90 min after drinking show dominant macrographism, elongation and separation. Similar patterns of these data might be very interesting in considering the reproducibility of SDT in spite of drinking alcohol. Data obtained at 105 min after drinking shows remarkable reduction from these macrographisms, elongation and separation. Thus, the data resembles those of the predrinking period (fig. 4).

The results of the data of all 17 subjects of the drinking group versus the results of the 30 subjects in the nondrinking group are as follows: Comparison of the whole length between the drinking and the nondrinking groups is made (fig. 5). The whole length of the drinking group in the predrinking state remains within the normal range as is usual among healthy adults. The value of whole length at 15 min after

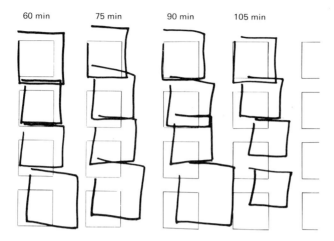

Fig. 4. STD done every 15 min after drinking. Note the macrographism after whisky drinking remaining for about 2 h.

drinking shows a definite increase to 167 mm (predrinking average 137 mm). Hereafter the increased value remains at an average of 165 mm for the following 2 h. The whole length for the nondrinking group shows a gradual increase which is perhaps due to physical and mental fatigue produced by frequent trials of the test. However, it is obvious that the elongation in the drinking group from 15 to 75 min after drinking is caused by the influence of alcohol to the vestibulo-cerebello-cerebral-spinal system.

Another explanation of the parameters is the change in each length of every square (R-1 to R-4). Most healthy adults draw each line of one square slightly shorter than the 40 mm defined on the test paper. On the contrary, the separation (distance between the start and the end) becomes larger with the lower drawing. Therefore, the largest separation is usually seen in the R-4 square. In the drinking group, the value of length of each line of a square exceeded 40 mm for the most part. The difference between these values is seen as the largest in R-4. These facts might indicate that the drinking subjects could no longer draw exactly along the square because of disturbed function of the vestibulospinal system due to alcohol affection.

Meanwhile, study of stabilography on standing upright showed some prolongation in locus of the gravity as influenced by taking alco-

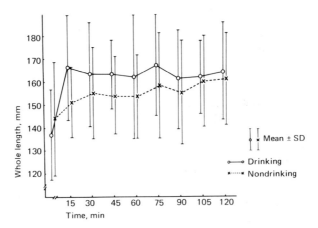

Fig. 5. Change of 'whole length', which is one of the parameters of the SDT, at the predrinking stage and during the following 2 h of alcohol consumption; comparison of the means and SD of measured whole length between the drinkers and nondrinkers. Note the marked increase of the means of whole length in the drinkers to that of nondrinkers.

hol, and this tendency was noted after drinking for 15 or 30 min. Considering the results of the tests for the vestibulo-ocular system, no spontaneous nystagmus was observed in either the drinking or the nondrinking group at any period during the test. Even in the eye tracking test no abnormal finding was observed except for one case, who after drinking for 15 min, showed a weak saccadic pursuit.

Since many findings of the examinations mentioned above, such as (1) positive findings of SDT (vestibulospinal system, arm), (2) no notable affection on the standing test (vestibulospinal system, lower legs), (3) no detectable spontaneous nystagmus, and (4) none of the notable influences on the eye tracking were neurootologically considered, there are some reasonable speculations that alcohol might strongly influence a higher center in the midbrain, releasing the suppression effect on the lower motor neuron or reticular formation which is mainly related to motion activity to yield some excessive movements of the arm and hand, as macrographism in SDT.

H. Ohyama, MD, Department of Otolaryngology, Yamaguchi University, School of Medicine, Ube City 755 (Japan)

Adv. Oto-Rhino-Laryng., vol. 30, pp. 311–314 (Karger, Basel 1983)

Evaluation of Ataxia by Measuring Changes in Angulation of Shoulders while Stepping

K. Taguchi, T. Ishiyama, M. Kikukawa, H. Yachiyama, K. Higaki and C. Hirabayashi

Department of Otolaryngology, Shinshu University School of Medicine, Matsumoto, Japan

Ataxia results in a limited ability to coordinate voluntary muscular movements while stepping or walking due to the loss or disturbance of body balance organs.

This study deals with the change of shoulder angulation and its variance in normal persons and subjects with several peripheral vestibular disorders in order to evaluate ataxia.

Material and Method

Subjects

Normal subjects were chosen from normal adults between 18 and 30 years of age without any ear disease or locomotor disorder. Subjects with vestibular disorders were 10 patients with Ménière's disease, 2 with vestibular neuronitis, 4 with sudden deafness accompanied by vestibular insufficiency and 5 with other peripheral vestibular disorders.

Apparatus

Changes in angulation of shoulders while stepping were measured using a polarized light goniometer. Light from the projector was passed through a modulator to provide a rotating beam of phase-polarized light the axis of which was perpendicular to the frontal plane of the subject. Photoelectric sensors were fitted to the subject's shoulders and the position of the body's center of gravity. The phase differences between two sensors were fed into a data recorder which separately stored the data of the bilateral shoulder angulations with respect to the center of gravity on the magnetic tape. The recorded data on the tape were fed into a computer to calculate the real value of change in angulation.

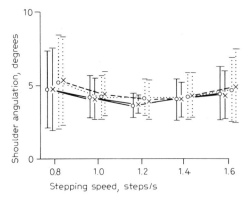

Fig. 1. Shoulder angulation while stepping. Mean values($\pm$ SD) of real shoulder angulation with respect to the center of gravity. $\bigcirc-\bigcirc$ = eyes open, right shoulder; $\times - \times$ = eyes open, left shoulder; $\bigcirc \cdots \bigcirc$ = eyes closed, right shoulder; $\times --- \times$ = eyes closed, left shoulder.

Results

Normal Subjects

Figure 1 shows the real values of changes in shoulder angulation during stepping movements in normal subjects. No statistically significant differences were shown between the values with eyes open and those with eyes closed in both shoulders ($p > 0.05$).

Subjects with Peripheral Vestibular Disorders

The patients with peripheral vestibular disorders other than vestibular neuronitis showed characteristic results in which the shoulder of the same side with affected ear presented larger values and wider distributions (table I).

Table II shows maximum and minimum values which had been obtained from 2 patients on the time course of diseases in 1 year. The patients with Ménière's disease presented a remarkable change depending on the stages of the disease.

Discussion

Stepping is a basic movement of walking. So far many recording techniques of walking have been devised. Recently a polarized light go-

Table I. Real shoulder angulation while stepping in patients with peripheral vestibular disorders (1.2 steps/s, means ± SD, degrees)

Kinds of lesions	Ménière's disease		Vestibular neuronitis	Sudden deafness		Others	
Side of lesion	R	L	R	R	L	R	L
Number of cases	5	5	2	2	2	2	3
Eyes open							
Right shoulder	9.6 ± 4.1	5.7 ± 2.4	3.8 ± 1.0	6.8 ± 1.8	3.4 ± 1.0	3.9 ± 1.1	4.1 ± 1.0
Left shoulder	5.8 ± 2.5	9.8 ± 3.9	3.6 ± 0.9	4.2 ± 1.5	5.2 ± 1.7	3.8 ± 0.8	5.0 ± 2.2
Eyes closed							
Right shoulder	10.0 ± 4.7	6.2 ± 3.0	4.0 ± 1.2	7.4 ± 2.8	4.7 ± 2.0	6.6 ± 1.8	4.2 ± 1.6
Left shoulder	7.4 ± 3.5	9.8 ± 4.5	4.0 ± 1.5	5.8 ± 1.9	6.1 ± 2.0	4.4 ± 1.5	6.7 ± 1.9

Table II. Time course of real shoulder angulation (1.2 steps/s, means ± SD, degrees)

Cases			Maximum	Minimum
Right Ménière's disease	Eyes open	Right shoulder	11.3 ± 5.5	6.4 ± 3.1
(67-year-old male)		Left shoulder	8.0 ± 4.1	4.3 ± 1.9
	Eyes closed	Right shoulder	13.1 ± 8.1	8.1 ± 3.7
		Left shoulder	7.6 ± 4.6	5.0 ± 2.3
Right vestibular neuronitis	Eyes open	Right shoulder	4.0 ± 1.4	3.6 ± 1.0
(19-year-old female)		Left shoulder	3.9 ± 1.0	3.6 ± 1.1
	Eyes closed	Right shoulder	5.1 ± 2.1	4.4 ± 1.6
		Left shoulder	4.8 ± 1.9	4.1 ± 1.2

niometer has been devised and used for analysis of walking [*Grieve,* 1969]. The goniometer can record a rotating movement which had been difficult to obtain by other methods. The technical defect of the goniometer is its inability to record phenomena when something intercepts the light from the projector illuminating sensors [*Taguchi and Kikukawa,* 1982].

In the previous study we obtained patterns of shoulder angulation and normal values of change in shoulder angulation. These patterns and values are not real angulation, but they consist of the change in shoulder angulation, the angle of body rotation around its vertical axis and the disparity between bilateral shoulder movements. We devised a method of calculating the real change in shoulder angulation while stepping, using a minicomputer system. A close relationship is shown between the results obtained manually from angle diagrams and real values obtained by computer calculation. However, the measurement of real values is more useful to evaluate ataxia and to determine the affected side of some vestibular disorders.

References

Grieve, D.W.: The assessment of gait. Physiotherapy, Lond. *56:* 452–460 (1969).
Taguchi, K.; Kikukawa, M.: Measurement of changes in angulation of shoulders during stepping movement. Practica otol., Kyoto *75:* 290–296 (1982).

K. Taguchi, MD, Department of Otolaryngology, Shinshu University School of Medicine, Matsumoto 390 (Japan)

Adv. Oto-Rhino-Laryng., vol. 30, pp. 315–318 (Karger, Basel 1983)

Influence of Head Tilting on Maintaining the Arms in the Horizontal Position

Tatsuya Okuzono, Toru Sekitani, Masutoshi Nishikawa,
Masaaki Hiyoshi, Tetsuyasu Hirata, Toshishige Kido

Department of Otolaryngology, Yamaguchi University, Ube, Japan

The involuntary ability of the hand-arm to keep an object, e.g. a cup of water, in the horizontal position, is a well-established balance reflex of the hand-arm and body. The purpose of this study is to assess the effect of voluntary tilting the head on arm and body balance.

Methods

The subject stands on a detecting platform and a strain-gauge-type linear accelerometer is attached to his head, while he grasps with both hands, a bar to which accelerometers used for the detection of arm balance were attached (fig. 1). The output from three detectors (hands, head and platform) with an X-Y direction was digitalized by a homemade A/D converter, then stored on a floppy disk and analyzed by microcomputer. The subjects were 30 healthy persons without equilibrium disturbances and some patients with vertigo.

During the examination, the subject was required to stand upright on the platform with his feet together and arms stretched forward. While the subject maintained this position, the following seven-step examination was performed: (1) Past pointing test position with eyes open, then (2) with eyes closed; (3) with head tilted 40° to the right with eyes closed; (4) return to the first position with eyes closed; (5) with the head tilted 40° to the left with eyes closed; (6) return to the first position with eyes closed, and finally (7) keeping the first position with eyes open. The duration of each step was 10. Arm balance was measured as the extent of deviation in one direction occurring at each step, and body balance was evaluated using the 'total sway': a total amount of body sway, of each stabilogram, and statokinesigram (SKG) with its vector analysis.

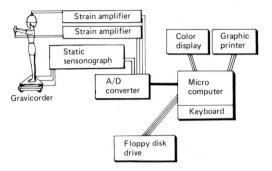

Fig. 1. Apparatus for head, arm and body balance recording. The output from three detectors (hands, head and platform) with an X-Y direction was digitalized by a home-made A/D converter, then stored on a floppy disk and analyzed by microcomputer.

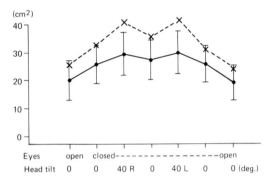

Fig. 2. Note difference in total sway between the averaged values of normal subjects and that of the patient (×) in the head-tilted position.

Results

From the typical recording of a normal subject, we could find the following. When he closed his eyes, the movement of his body increased, but his arm balance was almost even. Upon tilting his head with his eyes closed, the arm balance showed some changes in stability: his arm tilted slightly to the same direction as his head for a short period of time just after he tilted of his head, but the arm balance soon recovered its former stability. Some subjects showed good ability in keeping their arm balanced and the accelerometer bar vertical. In the stabilogram no notable increase in body sway was observed with tilting of the head, but the center of gravity seemed to move forward.

Healthy subject

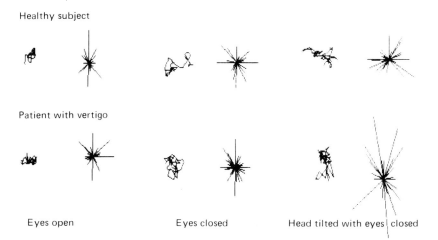

Patient with vertigo

Eyes open   Eyes closed   Head tilted with eyes closed

Fig. 3. Statokinesigram and its vector analysis. There is no notable increase in body sway either with eyes closed or with the head tilted in healthy subject (upper line). On the contrary, in the patient with vertigo (lower line), the body sway increased, especially with head tilting.

From the recording obtained from a patient with vertigo due to basilar insufficiency, we could find that body movement and arm imbalance increased, not only with his eyes closed, but with his head tilted as well. He felt so unsteady that he could not tilt his head the full 40° for fear of falling over. Figure 2 shows the average 'total sway' and standard deviation from the test results of 30 healthy subjects. It reveals a relatively notable difference in the body balance both with eyes open and with eyes closed, but no notable difference between the normal and head-tilted positions. ×----× is the recording obtained from the patient with vertigo, compared to the average 'total sway' and standard deviation of the normal subjects. Please note the difference in the 'total sway' between the normal subjects and the patient in the head-tilted position. Figure 3 shows SKG and its vector analysis (V-SKG) of the normal subject and the patient with eyes open, eyes closed and head tilted with eyes closed. There is no notable increase in body sway either with eyes closed, or with the headtilted position in healthy subjects (upper line). On the contrary, in the patient with vertigo (lower line), the body sway increased especially with head tilting. In addition, V-SKG revealed a remarkable increase in body sway and direction, which could not be clarified by SKG alone.

Discussion

In normal subjects, the effect of head tilting is not as strong as that of eye closing on arm and body balance because the deviation produced by head tilting was quickly compensated for by the 'righting reflex'. Therefore, the subjects could keep an upright position without a large amount of body sway or losing their balance. On the contrary, in patients with vertigo the effect of head tilting was stronger than that of eye closing on the arm and body balance because the deviation produced by head-tilting could not be compensated for by the righting reflex: the patients exhibited a poor ability in maintaining equilibrium and almost lost their balance. Therefore, the patients had to struggle to keep upright more than the normal subjects.

There are three significant factors which cause difficulties in maintaining coordination in the voluntary head-tilted, eyes-closed position. These factors are: (1) a change in head position; (2) a change in center of gravity, and (3) a change in muscle tonicity in the neck, arms, feet and other parts of the body. These changes caused some deviation in arm and body balance, deviations which could not be compensated for by the 'righting reflex' alone. Most patients regained their balance only after returning their head to the normal position or after opening their eyes.

Conclusions

To check the stabilizing ability of arm and body, we tried to analyze the recordings. From the recordings of arm and body balance of normal subjects and patients in the voluntary head-tilted position with eyes closed, we concluded the following: In normal subjects, the effect of head tilting was not as strong as that of eye closing on arm and body balance. In the patients with vertigo, the effect of head tilting was stronger than that of eye closing on arm and body balance. Therefore, the head-tilting test is one of many significant tests for the evaluation of the 'righting reflex'. From the test results, we have obtained in our qualitative tests thus far, we are encouraged and intend to research this phenomenon further with quantitative testing.

T. Okuzono, MD, Department of Otolaryngology, Yamaguchi University School of Medicine, Ube City 755 (Japan)

Adv. Oto-Rhino-Laryng., vol. 30, pp. 319–329 (Karger, Basel 1983)

Panel Discussion Synthesis:
Neurophysiological and Diagnostic Aspects of
Vestibular Compensation

W. Precht[a]
Panelists: H. Flohr, M. Lacour, K. Kaga, W. Precht, C.R. Pfaltz
[a] Institute for Brain Research, University of Zürich, Switzerland

The aim of the panel was: (1) to discuss briefly the most recent advances in the understanding of the mechanisms leading to functional recovery after unilateral labyrinthectomy or vestibular neurotomy as derived from animal experiments, and (2) to evaluate diagnostic procedures allowing the clinicians to define the state of the vestibular system following various time intervals after such lesions.

Over the past 100 years many researches were attracted by the remarkable functional recovery (often referred to as vestibular compensation) occurring after unilateral vestibular lesions in many species. The vestibular lesion model was and is, therefore, frequently being used as an experimental paradigm to study motor learning and the plastic, adaptive mechanisms associated with it.

Apparently, in the vestibular system the bilateral complementary inputs are essential for proper function and survival and when disturbed the CNS makes every effort to reestablish function. As we shall see this process is not a simple one and involves the cooperative efforts of at least three sensory systems, the vestibular, optokinetic and proprioceptive systems [for review of the classical literature see *Schaefer and Meyer*, 1974, and *Precht*, 1974].

The recovery process has to deal with two major tasks: (1) it has to reestablish *balance*, i.e. reduce or abolish the asymmetry in postural

tone of body and eye musculature thereby causing head and body tilt
and deviation as well as ocular nystagmus to vanish; (2) the gain of dynamic vestibular reflexes has to be recalibrated to assure symmetrical
compensatory vestibulospinal (VSR) and vestibuloocular (VOR) reflex
action during motion of body and head.

With the above in mind the panel dealt with both balance and gain
control after lesion in VSR and VOR. The VSR were discussed first,
and *Flohr* summarized his findings and views on the mechanisms leading to rebalance of head posture in hemilabyrinthectomized frogs as
follows.

The fact that recovery of head posture occurs in the absence of regeneration in the vestibular end organ or nerve necessarily implies an
extensive reorganization of the remaining structures involved in the
control of posture. The nature of the central changes leading to functional revovery is as yet little understood. The attempts to understand
these processes seem to concentrate mainly on three points: (1) the
identification of the basic cellular mechanisms underlying the compensation process. Probably this involves *synaptic* changes and this approach therefore consists in the search for synaptic modifications that
occur during the compensation process; (2) the *localization* of the
changes responsible for compensation. As *Magnus* [1924] – and more
recently *Llinás and Walton* [1979] – emphasized, the compensation
process probably results in the formation of a new and complex circuitry. It is probably the product of multifocal changes that implies a
broad distribution of functional modifications; (3) the characterization
of factors and mechanisms that *control* the adaptive reorganization.
This reorganization involves the rewiring of diverse, widely separated
parts of a damaged system. And this rewiring does result in new meaningful circuits that are adapted to functional requirements.

It has been shown that compensation is a goal directed process, induced by the system's functional error and directed to its elimination.
The asymmetrical input resulting from hemilabyrinthectomy can be
systematically modified by subsequently exposing the experimental
animal to *centrifugation,* which alters the vector of the summed acceler-
ative and gravitational forces acting on the otoliths [*Flohr* et al., 1981].
An *increase* of the g-forces acting on the otoliths of the hemilabyrinthectomized animal causes a considerable acceleration of the compensation. Exposing the hemilabyrinthectomized animal to decreased g-forces prevents the onset of the compensation if started immediately after

hemilabyrinthectomy as long as the treatment is continued. It stops the compensation process and reverses the already attained compensation if started later.

It follows from the above that the initiation, maintenance and formation rate of the neural modifications responsible for compensation depend on the error signal (i.e. the asymmetrical input signal). It is therefore improbable that vestibular compensation is a direct consequence of the lesion induced, for example, by chemical factors generated in the vicinity of the lesion. Rather it appears to be the result of *specific control mechanisms* that guide adaptive changes, including recovery processes in the adult CNS. This means that neural nets capable of learning and recovery in addition to having modifiable components also have mechanisms for the regulation of such changes.

The Marr-Albus-Ito hypothesis on the adaptive modification of the VOR following long-term optical reversal of vision proposes a feedback control of synaptic connectivity within the VOR system similar to that suggested above. The crucial element in this hypothesis is a heterosynaptic modification of specific synapses, e.g. at the Purkinje cell synaptic level.

In this context our recent observations that suggest that *neuropeptides* are involved in vestibular compensation might be relevant: (1) the compensation process can be slowed down by hypophysectomy; (2) the compensation process so impaired can be restored by the administration of $ACTH_{4-10}$, which is a fragment of the ACTH molecule devoid of the hormonal activity of ACTH; (3) $ACTH_{4-10}$ treatment considerably accelerates the compensation process, and (4) the specific $ACTH_{4-10}$ antagonist $[D-Phc^7]-ACTH_{4-10}$, which contains a dextrorotatory amino acid in position 7, inhibits vestibular compensation.

These findings suggest that $ACTH_{4-10}$-like neuropeptides might be physiologically involved in compensation processes. As an initial conjecture it is suggested that they might be part of the above proposed control mechanisms that direct adaptive modifications in the adult CNS.

The logistic structure of such a humoral system would be different from that implicated by the concept of heterosynaptic modification. The cellular effects of such humoral factors would be anatomically generalized rather than localized. This diffuse broadcast, however, is exactly what is needed for bringing about goal-directed, concerted modulations in a distributed system, eventually producing a symmetrical neural output in descending spinal pathways controlling neck, trunk and limb musculature in the resting pose of the animal.

Lacour discussed the dynamic aspects of VSR during the postlesion recovery period as derived from studies performed in monkeys and cats. He assumed that vestibular compensation needs the functional integration of multisensory inputs which tonically and dynamically substitute for the missing labyrinthine afferents. He therefore analyzed the role of the remaining labyrinthine afferents, of somatosensory inputs and of visual information in the recovery of both static and dynamic motor functions in unilateral vestibular neurectomized animals. His main results and the conclusions were as follows: (1) All the sensory inputs involved in perception of space, body posture and body motion in space intervene in the recovery process. (2) Vestibular compensation develops in active animals only, i.e. when they can use all available information elicited by an active sensorimotor exploration. In this multisensory substitution process, the remaining labyrinth appears particularly involved in recreating the dynamic counterpart. (3) The earlier all information is provided, the better adapted the recalibration of gain will be and the better the final compensation. This point was confirmed in human patients submitted to a unilateral vestibular neurectomy: they showed a faster recovery in the presence of sensorimotor activity during the early stages of compensation. These findings suggest that a 'sensitive period' probably exists for functional recovery, which may be crucial for achieving functionally well- or maladapted behavior. He concluded that vestibular compensation is achieved by means of a multisensory substitution process requiring an actively behaving subject. Vestibular compensation resembles a sensorimotor relearning process implicating the concerted activity of many integrative central nervous structures. This could explain why any defect in any sensory input or central nervous structure delays the recovery when occurring during the early stages of compensation. This concept does not exclude the role of plastic or functional changes at the cellular level. Neural plasticity may have a complementary role in adult subjects. From a clinical point of view, he believed that these results may lead to new training programs to improve the recovery.

Kaga extended the discussion of static and dynamic VSR to man and evaluated the importance of VSR testing in the diagnosis of vestibular malfunction as well as in follow-up study of recovery of function.

In dealing with vestibular compensation in man one should realize that the lesion often develops slowly, may not affect all endorgans and may involve other structures indirectly such as the flocculus in the case of tumors originating from the ganglion Scarpae. At the time when labyrinthectomy or neurotomy is performed the CNS, therefore, probably is in a state quite different from that found in experimental studies, i.e. compensatory processes may be already at work at the time of surgery. Thus, a direct comparison of experimental and clinical studies is quite difficult.

A large test battery is available to diagnose static and dynamic VSR malfunction. According to *Kaga, static* VSR testing includes the Romberg and Mann test, standing on one foot test, the tilting board test and labyrinthine righting reflexes, whereas *dynamic* testing implies stepping past-pointing test and vertical writing test, arm tonus reaction and tandem gait. A quantitative test involves measurement of changes of center of gravity in standing subjects.

It is clear that various of these tests measure both static and dynamic VSR and that separation of the two is not always an easy task. *Kaga* presented work done with 8 patients who underwent unilateral labyrinthectomy and were studied 2–8 years postoperatively. All had suffered from severe vertigo and were almost completely deaf on the affected side. The fact that they showed imbalance symptoms after surgery, however, indicated that the periphery was partially intact. As to be expected VSR recovery was better when tested in the light. In the dark, stepping tests revealed body deviation to the lesioned side which subsided within 2–4 months after surgery. Subjective instability lasted anywhere from 3 to 12 months. In one case the Mann test was abnormal 8 years postoperatively while all other tests appeared normal. Measuring the center of gravity in Ménière patients undergoing unilateral labyrinthectomy revealed that recovery was perfect within 1 month when measured in light but required 6–12 months when measured in the dark. These cases may come closest to experimental conditions. The findings indicate the important role of other sensory systems for VSR recovery.

The panel now turned to the VOR; experimental studies were discussed by *Precht* and clinical aspects summarized by *Pfaltz. Precht* referred mainly to work done with higher mammals and summarized the major recent findings obtained in his laboratory using the cat as an experimental animal [*Maioli* et al., 1983; *Ried* et al., 1983].

Following lesion of one vestibular nerve both the balance and gain control systems serving the VOR are disturbed, i.e. in most animals there exists, at rest, a strong nystagmus and, during head rotation, an impairment of VOR dynamics.

The spontaneous nystagmus is strongly reduced within the first week even when measured in the dark. A slight imbalance, i.e. an eye drift of some 3°/s, often towards the lesioned side may, however, persist for periods of over 1 year but only when the animal is in the dark. The eyes seem to drift towards a new null-point which is shifted towards the lesioned side and may stay there if no saccades occur. When the animal makes saccades to one or the other side of that null-point the eyes drift back towards that point, i.e. gaze holding failure is apparent. Thus, the drift may, in a strict sense, not be identical to spontaneous nystagmus observed immediately after the lesion.

There is experimental evidence suggesting that fast reduction of spontaneous nystagmus is produced by the reestablishment of a balanced resting firing rate in the bilateral vestibular nuclei as well as in other structures such as the cerebellum and the inferior olive [*Precht* et al., 1966; *Llinás and Walton*, 1979]. Balance in neuronal resting rate may be driven by ascending and descending inputs and represents an effort of the 'whole' CNS towards balance rather than a process unique to one subsystem.

Contrary to the rather fast occurrence of oculomotor balance the dynamic VOR deficiencies persist much longer. Acutely after the lesion the gain of the horizontal VOR induced by rotation to the intact side in the dark falls by ca. 50%. The VOR accompanying opposite rotation drops even more, thereby producing a highly asymmetric response. When the spontaneous nystagmus has subsided, VOR asymmetry at low gain levels persists although some cats show an increase in VOR gain already within the first 10 postoperative days. Those animals that do not show this early gain increase exhibit a gradual and slight improvement of gain developing over many months and eventually show some improvement in VOR symmetry. Interestingly, when the lesion was made at young age (6 weeks) symmetry and gain improved more readily.

The long persistence of asymmetry in most animals is difficult to understand in view of the hypothesis that the resting rate in the bilateral vestibular nuclei is reestablished in parallel with balance control. Bilaterally symmetrical resting rate should lead to a symmetrical VOR as has actually been demonstrated by unilateral canal plugging [*Zuck-*

erman, 1967]. Either balance of vestibular resting rate is not the reason for the rebalance of VOR or, if it is, we have to assume that units show rearrangement of synaptic circuitry.

Recent single unit work in the cat, indeed, suggests that in the de-afferented vestibular nuclei much fewer units responded to rotation when compared to those of the intact side. This difference in bilateral responsiveness of vestibular units could account for some of the persisting VOR asymmetry.

In addition to gain and symmetry VOR phase likewise showed persisting deficits consisting of a parallel phase shift towards larger lags at all frequencies and to both directions. The explanation for this change is not clear at present.

Interestingly, these long-lasting deficiencies in VOR performance are also accompanied by chronic deficits in the OKN system. Acutely after the lesion the OKN produced by pattern movement to the intact side was very poor compared to that produced by opposite motion. This deficiency persisted to various degrees in most animals studied for periods of many months. In addition, OKAN occurring after stimulation towards the intact side was permanently abolished. As with persisting vestibular deficits OKN abnormalities may be explained by the poor population response in the vestibular neurons on the lesioned side (see above).

In spite of all these long-term deficiencies in the VOR/OKN system after unilateral vestibular lesions it is quite surprising to observe an apparently normal locomotor behavior of the animals. In fact, it is very difficult to identify lesioned animals in a group containing also intact animals. Given, however, that the VOR in the freely moving animal is assisted by optokinetic as well as neck proprioceptive afferents and that particularly the latter show plastic adaptive modifications in gain, it is not surprising that gaze is stabilized. The advantage of multisensory control of VOR reflex behavior is evident. It may well be that these other inputs together with the impaired vestibular system are sufficient in an overall functional sense so that the adaptive vestibular gain control system is only weakly activated. That it can operate in this condition is evidenced by the occurrence of early gain increases in some animals and slight gradual gain changes in others.

Finally, it should be pointed out that the above experimental work in the cat was based on vestibular neurotomies including the ganglion Scarpae. As it turned out in the discussion it is not clear whether pure receptor lesions (labyrinthectomies) with minimal involvement of primary afferents yield a similar picture. *Igarashi* found that peripheral lesions in the monkey did not cause major degenerations in ganglion cells. Thus, in this condition central terminal degeneration is expected to be absent and central reorganization may be different. Also ganglion cells may be providing some resting discharge that would contribute to compensation. All these points, however, require experimental verification.

In conclusion the experimental studies indicate that following lesion of one vestibular nerve several vestibular and also optokinetic functions remain impaired when tested in the dark but are assisted effectively by multisensory inputs when the head is free to move in a lit environment. Some adult cats and particularly young animals showed an amazing recovery even when tested in the dark. This recovery occurred fast, i.e. it started about at the time when spontaneous nystagmus had disappeared (day 4–5) and was accomplished by day 10. As already pointed out by *Lacour* (see above) there may exist an early critical period during which these changes can occur. Possibly, those animals that improved gain may have been more active during this critical period thereby inducing adaptive changes. Those that missed this chance showed only a very protracted and smaller recovery of VOR gain. These findings may suggest the importance of early training for recovery.

Pfaltz discussed the clinical aspects of VOR recovery and included OKR testing results to emphasize their diagnostic value. According to *Pfaltz,* for the clinician one of the most important problems concerns the reliable criteria indicating the achievement of vestibular compensation and in conjunction with it the assessment of working capacity.

In a previous study, *Pfaltz* and associates had followed up a series of 100 head injuries, who were submitted to repeated neuro-otological examinations [*Meran* et al., 1978]. The results were evaluated with respect to the diagnostic importance of vestibular symptoms and the OKN. Positioning and positional nystagmus were found to be the most common symptoms, indicating a vestibular lesion which was not yet compensated. They occurred in 51% of the injured individuals. Spontaneous nystagmus could only be demonstrated in 25% of the cases although all the patients were suffering from posttraumatic vertigo. Pathologic caloric responses occurred only in 34% of the cases and abnormal OK responses were even more rare (9%).

In a group of 100 cases of sudden unilateral loss of vestibular function, followed up over a longer period of time (1969–1979) the following observations were made [*Pfaltz and Meran, 1981*]: Vestibular symptomatology is predominantly uniform and characteristic but depends mostly on the site of the lesion. Vestibular symptoms and the great variability of the course of vestibular compensation indicate that there is no constant localization of the vestibular lesion. Prognosis of functional restoration, either by recovery or by central compensation, depends primarily on the pathogenesis and the site of the lesion and in the second place on the age and the general functional state of the patient's central nervous system. Cerebral arteriosclerosis and senile degenerative encephalopathy involving the brainstem, generally cause a substantial delay of either spontaneous recovery or central compensation of vestibular lesions, both in the head injury and in the vestibular paralysis group.

In two recent studies, carried out in 50 normal subjects and 100 patients affected with well-defined peripheral and central vestibular disorders, the modification of OKR responses by peripheral and central lesions of the vestibular system were investigated [*Pfaltz and Ildiz, 1982*]. The results may be summarized as follows. From a merely statistical point of view investigation of foveoretinal OKN is a far more reliable test procedure than that of foveal OKN. Its responses show little interindividual variations, probably because this type of OK response does not primarily depend on the test person's attentiveness and cooperation, but first of all on the integrity of the reflex mechanism.

In *peripheral vestibular lesions* gain is normal, whereupon in patients with a spontaneous vestibular nystagmus ipsidirectional saccades will occur as well as a corresponding directional preponderance (DP). This DP is more conspicuous at slow target velocity ($< 30° \mathrm{s}^{-1}$). In *central lesions* distortion of the slow component of OKN by saccades in both directions, slow build-up of foveoretinal OKN, and low gain of OKN, decreasing with rising target speed were observed. Since *foveoretinal OKN* is strongly influenced by the functional state of the vestibular centers it is one of the most reliable tests with respect to the differential diagnosis between a peripheral and a central vestibular disorder. From a *topodiagnostic point of view* the OK test is more important than vestibular tests based on the experimental investigation of the VOR. On the other hand, *not OK* but only *VOR responses* reflect the state of vestibular compensation. *Achievement of recovery or central compensation* of vestibular disorders may, therefore, only be assessed by specific vestibular test procedures. With respect to the assessment of vestibular compensation it should be emphasized that spontaneous nystagmus or DP of experimentally induced vestibular nystagmus as single symptoms may not be overrated because they may persist unnoticed for years without any functional disturbances. However, provoked, i.e. positional and positioning nystagmus, particularly the true paroxysmal type, as a rule indicate an uncompensated vestibular lesion. These symptoms are usually accompanied by vertigo. With respect to the *assessment of vestibular compensation* clinical importance of provoked vestibular nystagmus is much greater than that of spontaneous or experimentally induced vestibular nystagmus.

Based upon the results of numerous experimental studies [*McCabe* et al., 1972] and correlating these results with clinical observations in man, *Pfaltz* suggested that the process of vestibular compensation may be subdivided as follows:

Stage 1: Acute phase of partial or total loss of vestibular function: spontaneous nystagmus to the normal ear. Positional/positioning nystagmus to the normal ear, accompanied by pathologic VSR (severe – ataxia).

Stage 2: Accommodation (specific): progressive suppression of spontaneous and positional/positioning nystagmus, pathologic VSR (mild – overcompensation).

Stage 3: Central multisensory compensation (nonspecific): by visual and somatosensory mechanisms suppression of spontaneous and provoked nystagmus. Suppression of pathological VSR.

Stage 4: Recovery and overcompensation: spontaneous nystagmus towards lesion (recovery nystagmus). Positioning nystagmus towards lesion. Normal VSR.

It is hoped that this attempt to combine experimental and clinical views of vestibular compensation will stimulate discussion of the topic among the Bárány society members and lead to mutual understanding of problems and concepts. Clearly these are many similarities as well as differences between clinical and experimental studies, partly because in the former the conditions at the time of the lesion are different from those in animals and also because there are differences in the normal reflex performance between man and experimental animals, particularly in the OKR which is so intimately associated with the vestibular system.

References

Flohr, H.; Bienhold, H.; Abeln, W.; Macskovics, I.: Concepts of vestibular compensation; in Flohr, Precht, Lesion-induced neuronal plasticity in sensorimotor systems, pp. 153–172 (Springer, Berlin 1981).

Llinás, R.; Walton, K.: Vestibular compensation: a distributed property of the central nervous system; in Asanuma, Wilson, Integration in the nervous system. (Igaku Shoin, Tokyo 1979).

Maioli, C.; Precht, W.; Ried, S.: Short and long-term modifications of vestibuloocular response dynamics following unilateral vestibular nerve lesions in the cat (in press).

Magnus, R.: Körperstellung (Springer, Berlin 1924).

McCabe, B.F.; Ryu, J.H.; Sekitani, T.: Further experiments on vestibular compensation. Laryngoscope 82: 381–396 (1972).

Meran, A.; Rohner, Y.; Pfaltz, C.R.: Zur Symptomatologie und Diagnostik der vestibulären Funktionsstörungen nach Schädelhirntrauma; in Kellerhals et al., Aktuelle Probleme der ORL, vol. 1, pp. 133–141 (Huber, Bern 1978).

Pfaltz, C.R.; Ildiz, F.: The optokinetic test: interaction of the vestibular and optokinetic system in normal subjects and patients with vestibular disorders. Archs Oto-Rhino-Lar. 234: 21–31 (1982).

Pfaltz, C.R.; Meran, A.: Sudden unilateral loss of vestibular function. Adv. Oto-Rhino-Laryng, vol. 27, pp. 159–167 (Karger, Basel 1981).

Precht, W.: Characteristics of vestibular neurons after acute and chronic labyrinthine destruction; in Kornhuber, Handbook of sensory physiology, vol. VI, Vestibular system, pp. 451–462 (Springer, Berlin 1974).

Precht, W.; Shimazu, H.; Markham, C.H.; A mechanism of central compensation of vestibular function following hemilabyrinthectomy. J. Neurophysiol. 29: 996–1010 (1966).

Ried, S.; Maioli, C.; Precht, W.: Vestibular nuclear neuron activity in chronically hemilabyrinthectomized cats (submitted for publication).

Schaefer, K.P.; Meyer, D.L.: Compensation of vestibular lesions; in Kornhuber, Handbook of sensory physiology, vol. VI, pp. 463–490 (Springer, Berlin 1974).

Zuckerman, H.: The physiological adaptation to unilateral semicircular canal inactivation. McGill med. J. 36: 8–13 (1967).

W. Precht, Institute for Brain Research, University of Zürich,
August-Forel-Strasse 1, Postfach, CH-8029 Zürich (Switzerland)

Adv. Oto-Rhino-Laryng., vol. 30, pp. 330–333 (Karger, Basel 1983)

Vestibular Compensation and the Significance of Rotation Tests

Marcel E. Norré

Department of Otoneurology and Equilibriometry, ENT Clinic,
University Hospital St. Rafaël, Leuven, Belgium

When monitoring patients with unilateral vestibular hypofunction, it is very important to assess the stage of compensation. The results of the rotation tests constitute a fundamental part of this evaluation.

For the follow-up of our patients we use a scheme which we present here. It is composed of three parts (fig. 1A): a basic follow-up scheme, a vestibulogram, and a comparative follow-up scheme. The following data are taken into consideration:

(1) *The complaints,* which are categorized according to the type of vertiginous sensation: 1 = spontaneous vertigo; 2 = provoked (positional) typical rotatory vertigo; 3 = atypical nonrotatory provoked vertigo; 4 = rather vague dizziness and unsteadiness; 5 = oscillopsia; 6 = no complaints.

(2) *The results of the caloric tests,* defining the degree of unilateral hypofunction, categorized as follows: unilateral areflexia; reaction at 14 °C; reaction at 20 °C; reaction at 30/44 °C with a difference of more than 50%, of 30–49%; less than 30% is considered normal.

(3) *The results of the rotation tests,* obtained by three types of rotation tests: acceleration of 3 °/s² during 30 s; postrotation reaction after a turning speed of 90 °/s stopped in 1 s; a sinusoidal rotation test – 0.05 Hz, five levels of intensity [*Norré, 1978*].

Directional preponderance of more than 20% is considered abnormal and categorized in two degrees: 20–29% ('20' in BFS, '2' in VG score: 1 point) and more than 30% ('30' in BFS, '2' in VG score: 2 points).

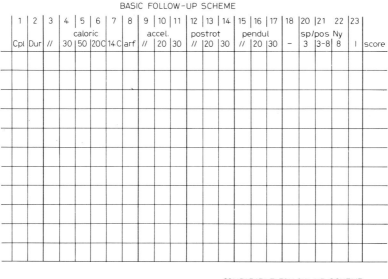

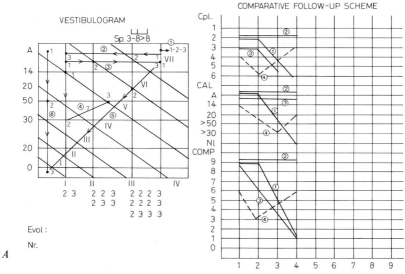

Fig. 1. The vestibulogram indicates the stage of gravity of vestibular dysfunction, independent of the complaints of the patient. We show some typical courses of the vestibulogram *(A)*: 1 = unchanged state of dysfunction; 2 = horizontal course with persisting unilateral hypofunction but improving results of the rotation tests; 3 = horizontal course in opposite direction; 4 = a fluctuant course; 5 = diagonal course with improvement of both caloric and rotation test results; 6 = vertical course with improving caloric hypofunction.

BASIC FOLLOW-UP SCHEME

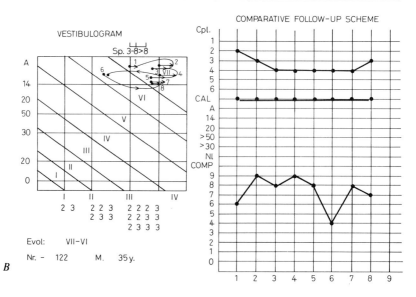

	1	2	3	4	5	6	7	8	9	10	11	12	13	14	15	16	17	18	20	21	22	23	
					caloric				accel.			postrot			pendul				sp/pos	Ny			
	Cpl	Dur	//	30	50	20C	14C	arf	//	20	30	//	20	30	//	20	30	–	3	3-8	8	I	score
1	2	3m						R	→			→			→				→				6
2	3	4m						R		→			→			→					→		9
3	4	6m						R		→		→				→					→		8
4	4	8m						R		→			→				→				→		9
5	4	11m						R		→			→		→						→		8
6	4	13m						R	→				→		//				→				4
7	4	20m						R		→			→		→						→		8
8	3	32m						R		→			→		→					→			7

For legend to figure 1 see reverse side.

(4) *The presence of spontaneous and/or positional nystagmus type II* (examination in 12 positions). The scoring is defined as follows: 1 point when nystagmus is present in 1–3 positions; 2 points for 3–8 positions; 3 points for more than 8 positions. The vestibular compensation is assessed by adding the scores of the data of (3) and (4). In this way we obtain a score of 0–9 points.

The *basic follow-up scheme* gives a survey of the several data in each examination and indicates side of hypofunction and direction of the directional preponderance (fig. 1b).

The *vestibulogram* gives a relationship between the degree of caloric hypofunction in the Y-axis and the degree of asymmetry in the three rotation tests in X-axis (degree I: asymmetry in only one rotation test; degrees II and III: asymmetry in resp. 2 and 3 rotation tests with the different combinations related to the degree of asymmetry of the reaction: 2 or 3.) Also the presence of spontaneous nystagmus is taken into consideration.

The *comparative follow-up scheme* shows the comparison between the course of the complaints, the caloric hypofunction and the stage of compensation assessed by the scoring system. We can see a parallel course (fig. 1a, ex. 1 and 2), an opposite evolution (ex. 3), fluctuations with parallel or opposite course (ex. 4). Using this scheme, we could study the follow-up of our patients. In this way we met various problems, among these we especially emphasize the persistent lack of vestibular compensation and the discordance between complaints and objective findings. Figure 1b gives an illustration of the first mentioned problem in a case with persistent unilateral vestibular loss, where no vestibular compensation developed according to the vestibular test results.

Reference

Norré, M.E.: The unilateral vestibular hypofunction. Acta Oto-Rhino-Lar. Belg. *32:* 421–668 (1978).

M.E. Norré, MD, Department of Otoneurology and Equilibriometry, ENT Clinic, University Hospital St. Rafaël, University of Leuven, B-3000 Leuven (Belgium)

Adv. Oto-Rhino-Laryng., vol. 30, pp. 334–337 (Karger, Basel 1983)

Response Characteristics of Lateral Vestibular Nucleus Neurons to Sinusoidal Tilt after Acute or Chronic Vestibular Deafferentation

C. Xerri, S. Gianni, D. Manzoni, O. Pompeiano

Istituto di Fisiologia Umana, Cattedra I, Università di Pisa, Italia

Unilateral labyrinth deafferentation produces a postural asymmetry, characterized by an extension of the limbs on the intact side and a flexion on the operated side [2, 5, 6]. These disturbances of body posture are also associated with deficits in locomotion and nystagmus. Both postural and dynamic deficits disappear to a large extent with time after the lesion. The asymmetric changes in posture and locomotion have been attributed to a severe unbalance in the spontaneous activity of vestibular nuclei neurons of both sides, which apparently decreases on the deafferented side, but increases on the intact side [5]. On the other hand, the compensation of the postural and motor deficits parallels the recovery of a symmetric and nearly normal spontaneous discharge in the vestibular nuclei neurons of both sides. Although the lateral vestibular nucleus (LVN) of Deiters exerts an important role in the postural regulation of the limb musculature as well as in locomotion [3], little is known about the effect of labyrinth deafferentation on the dynamic response characteristics of the corresponding neurons.

The present experiments were performed to investigate the response characteristics of LVN neurons to sinusoidal labyrinth stimulation in animals submitted to ipsilateral acute (aVN) or chronic vestibular neurectomy (cVN). The experiments were performed in decerebrate cats, paralyzed and artificially ventilated. The activity of 170 LVN neurons extracellularly recorded 2–3 days after ipsilateral aVN and 189 LVN neurons recorded 2–4 months after ipsilateral cVN was tested during sinusoidal rotation about the longitudinal axis of the whole animal at the standard parameters (0.026 Hz, ±10° peak amplitude) leading to stimulation of labyrinth receptors [1]. The unit activity was processed on-line with a computer system equipped with a fast Fourier analyzer. A spectral analysis of the angular input (table rotation) and of the output (unit activity) was performed and the gain (imp/s/°), sensitivity (% increase of the base discharge frequency per degree of peak displacement) and phase angle of the first harmonic of response were evaluated, the latter being expressed in arc degrees between the peak of the fundamental component of response and the peak of the side-down displacement of the animal indicated by 0°. The neurons were histologically located either in the rostroventral (rvLVN) or in the dorsocaudal part (dcLVN) of Deiters nucleus, which are known to project mainly to the cervical and the lumbosacral segments of the spinal cord, respectively [4]. The results obtained were then compared with those of 102 LVN neurons previously recorded in cats with the intact vestibular nerves [1].

After aVN, the proportion of responsive units decreased with respect to control experiments in rvLVN (from 91.2 to 72.2%) but not in dcLVN (from 67.6 to 69.9%). The mean discharge rate of the responsive LVN neurons decreased from 43.6 ± 54.4 (SD), imp/s in the control experiments, to 23.4 ± 20.7 (SD) imp/s. The average gain and sensitivity of responses of all the LVN neurons to tilt increased from 0.48 ± 0.44 to 0.61 ± 0.52 (SD) imp/s/°, and from 2.54 ± 2.71 to 3.50 ± 2.87 (SD) imp/s/° in % of base frequency, respectively. The average sensitivity (and to a lesser extent the gain) of responses of rvLVN neurons to the labyrinth input was almost twice that of the dcLVN units in preparations with the intact vestibular nerves; these regional differences disappeared after aVN, particularly due to an increase in gain and sensitivity of responses of dcLVN neurons. The proportion of LVN neurons, which were maximally excited by the direction of orientation of tilt, increased from 74% in the control to 86%. However, while in control experiments the proportion of units excited during side-down tilt (α-responses) was twice that of the units excited by side-up tilt (β-respones), after aVN the proportion of units showing a β-response increased particularly in dcLVN, due to reduction in number of units showing intermediate phase angle of responses. In addition, the average phase lead of responses relative to the extreme animal position decreased from $+12.3°$ in control experiments, to $+3.3°$.

After cVN, the proportion of responsive units was higher in rvLVN (76.7%) than in dcLVN (61.6%), thus approaching the values obtained in control experiments with the intact labyrinths. However, the mean discharge rate of the responsive LVN neurons (20.9 ± 17.1, SD, imp/s) did not change with respect to the values obtained after aVN, indicating that compensation of the postural deficits elicited by the vestibular neurectomy results from a redistribution of the excitatory drive within different populations of LVN neurons. The average gain and sensitivity of responses of the LVN neurons to the labyrinth input, which were particularly increased within the dcLVN after aVN, decreased after cVN, thus approaching the values obtained in preparations with the intact labyrinths. After cVN, the proportion of LVN neurons excited by the extreme animal position corresponded to 87.3%, whereas the number of units excited by side-up tilt (β-responses) was comparable to that of the units excited by side-down tilt (α-responses); these findings were similar to those obtained after aVN. However, the average phase lead of the responses relative to the extreme animal displacement increased from $+3.3°$ after aVN to $+16.2°$ after cVN, thus reaching the value obtained in preparations with the intact labyrinths.

The effects of aVN or cVN on the dynamic characteristics of responses of the LVN neurons to increasing amplitudes and frequencies of tilt were also investigated. Neither the gain nor the sensitivity of responses of the LVN neurons to tilt were modified in both types of preparations by increasing the peak amplitude of displacement from 5 to 20° at the standard frequency of 0.026 Hz. These findings are similar to those obtained in the control experiments and indicate that the linearity of the unit responses still persisted after ipsilateral labyrinth deafferentation.

Two main populations of LVN neurons were observed in control experiments by increasing the frequency of tilt from 0.008 to 0.325 Hz, $\pm 10°$ peak amplitude [1]. The first population of neurons usually exhibited a constant sensitivity and a maintained stability of the positional response (phase angle) over the frequency range. The second population, however, exhibited a progressive increase in sensitivity, which was associated in most instances with an increase in phase lag of the response for tilting frequencies higher than 0.026 Hz; these effects depend upon the dynamic responses of macular receptors to tilt [7]. After acute or chronic vestibular deafferentation the peak amplitude of the unit responses to stimulation of the contralateral labyrinth remained always in phase with the extreme animal displacement by increasing frequency of tilt. However, the increase in sensitivity of neuronal responses to tilt elicited in control experiments by increasing frequency of stimulation was suppressed after aVN, but partially recovered after cVN. The loss of dynamic sensitivity of LVN neurons after aVN may be responsible for the deficits in phasic postural reactions and the locomotor disorders described in these preparations. On the other hand the partial recovery of the dynamic sensitivity to tilt of chronically deafferented neurons indicates that the contralateral labyrinth plays an important role in the recovery of the dynamic equilibrium function.

References

1   Boyle, R.; Pompeiano, O.: Reciprocal responses to sinusoidal tilt of neurons in Deiters' nucleus and their dynamic characteristics. Arch. ital. Biol. *118:* 1–32 (1980).

2   Ewald, J.R.: Physiologische Untersuchungen über das Endorgan des Nervus oktavus (Bergmann, Wiesbaden 1892).

3   Pompeiano, O.: Vestibulo-spinal relationships; in Naunton, The vestibular system, pp. 147–180 (Academic Press, New York, 1975).

4   Pompeiano, O.; Brodal, A.: The origin of vestibulospinal fibers in the cat. An experimental-anatomical study, with comments on the descending medial longitudinal fasciculus. Arch. ital. Biol. *95:* 166–195 (1957).

5   Precht, W.: Characteristics of vestibular neurons after acute and chronic labyrinthine destruction; in Kornhuber, Handbook of sensory physiology, vol. VI/2. Vestibular system, part 2, Psychophysics, applied aspects and general interpretations. pp. 451–462 (Springer, Berlin 1974).

6   Schaefer, K.-H.; Meyer, D.L.: Compensation of vestibular lesions; in Kornhuber, Handbook of sensory physiology, vol. VI/2. Vestibular system, part 2, Psychophysics, applied aspects and general interpretations, pp. 463–490 (Springer, Berlin 1974).

7   Schor, R.H.; Miller, A.D.: Relationship of cat vestibular neurons to otolith-spinal reflexes. Exp. Brain Res. *47:* 137–144 (1982).

O. Pompeiano, MD, Istituto di Fisiologia Umana, Cattedra I, Università di Pisa, Via S. Zeno 31, I-56100 Pisa (Italy)

Adv. Oto-Rhino-Laryng., vol. 30, pp. 338–340 (Karger, Basel 1983)

Visual Compensation after Unilateral Sudden Loss of Vestibular Function

Kuniko (Uesugi) Furuya, Setsuko Takemori

Neurotology, Toranomon Hospital, Minato-Ku, Tokyo, Japan

The compensation after unilateral sudden loss of inner ear function is acquired by the central nervous system mechanism. It has been reported that vestibular type I and type II neurons were concerned with the compensation. Visual fixation is very important for maintenance of equilibrium. The body loses balance with eyes closed in cases of the acute stage of inner ear diseases and this is known as the Romberg phenomenon. The visual system assists equilibrium in cases of inner ear diseases. The purpose of this paper is to clarify the compensation acquired by the visual system after sudden loss of inner ear function.

Subjects

6 cases (1 male, 5 females), whose left labyrinth was operated, were used for this study. However, as 1 male case was infrequently examined, only 5 cases took part in this study. All cases were totally deaf in the left ear. The caloric responses of these cases were hypofunction in the left ear.

Methods

Operative Methods
The left labyrinth was totally destroyed under general anesthesia. Three bony semicircular canals were opened and the membranous labyrinth was destroyed. Streptomycin-sulfate powder was put into the operative space. The mastoid cavity was filled up by the temporal muscle.

Eye Movement Recordings

Eye movements were recorded by electro-oculography (EOG) before and after the labyrinthectomy. Spontaneous nystagmus was observed under Frenzel glasses and recorded with eyes open, closed and covered. The time constant was 3 s for eye movements and 0.03 s for eye velocity.

Optokinetic Nystagmus (OKN). A large optokinetic drum was rotated with a constant acceleration of $4°/s^2$ from 0 to a maximum velocity of $150°/s$, and decelerated from $4°/s^2$ to standstill. The drum was first rotated counterclockwise and then clockwise. The eye movements were recorded by EOG. The slow phase velocity of OKN which was evoked by $75°/s$ drum rotation was measured.

Eye Tracking Test. A round target with a diameter of 1 cm was moved horizontally. Eye movements and velocity were recorded by EOG. The small saccadic eye movements to both the right and left were counted during 10 cycles of eye tracking movements. The mean values were used.

Visual Suppression. 20 ml of cold water at $5°C$ was irrigated into the external auditory canal over a 20 s period in light and then the room lights were turned off. The patients were required to open their eyes in the darkness. After the slow phase velocity of caloric nystagmus had reached a constant and maximum response, the room lights were turned on for 5–10 s and the patients fixed their eyes on a target 50 cm above their eyes. After that the room lights were turned off again until the end of caloric nystagmus. The mean value of the slow phase velocity of caloric nystagmus in darkness (a) was measured. The mean value of the slow phase velocity of caloric nystagmus in light (b) was also measured. Using these two values, the percent of visual suppression was calculated as follows:

$$\text{Visual suppression (\%)} = \frac{a - b}{a} \times 100.$$

Results

(1) Spontaneous nystagmus lasted from 3 to 176 days after the operation. The spontaneous nystagmus recorded by EOG lasted longer than that observed by the Frenzel glasses.

(2) OKN: the asymmetry of OKN lasted from 13 to 183 days after the operation.

(3) Eye tracking test: the asymmetry of the numbers of saccadic eye movements to the right or to the left in the eye tracking test lasted from 13 to 183 days after the operation.

The asymmetry of OKN or the numbers of saccadic eye movements in the eye tracking test had disappeared before the spontaneous nystagmus disappeared.

(4) Visual suppression increased from 69 to 75–95% after the operation.

(5) Subjective findings: 1 case complained of unsteadiness even a year after the operation. However, 4 cases felt neither unsteadiness nor vertigo 30–47 days after the operation. At that time, spontaneous nystagmus was still observed or recorded by EOG. When the asymmetry of OKN patterns and eye tracking test have disappeared and visual suppression increases, the patients seem to recover from the unsteadiness.

Conclusion

The visual compensation after unilateral sudden loss of vestibular function can be examined by OKN, eye tracking test and visual suppression test.

K. Furuya, MD, Department of Otolaryngology, Nihon University, Faculty of Medicine, Kanda Surugadai 1–8–13, Chiyoda-Ku, Tokyo (Japan)

Adv. Oto-Rhino-Laryng., vol. 30, pp. 341–344 (Karger, Basel 1983)

Effect of Vestibular Stimulation on Bechterew Nystagmus

H.A.A. de Jong, M.M.J.M. Beukers, W.J. Oosterveld

Academisch Medisch Centrum, K.N.O.-Vestibulair onderzoek, Amsterdam,
The Netherlands

Introduction

When, in dogs, after a unilateral labyrinthectomy the other labyrinth is destructed, a reversal occurs of the compulsatory eye movement direction [1]. The fast phase of this so-called Bechterew nystagmus (BN) is directed to the side first operated upon. As both the labyrinths are destructed, the origin of a BN cannot be found in the peripheral vestibular system. In the present study the effect of linear accelerations on a BN in rabbits and pigeons was investigated.

Techniques

Labyrinthectomies in rabbits were done according to a known method [4, 5, 11]. In pigeons the membraneous inner ear structures were removed. The second labyrinthectomy was done 30 days after the first one. Vestibular tests were conducted 1 day after the second labyrinthectomy.

Results

In the ground based tests on the rotation chair and the parallel swing not any effect on the eyemovements was found. Position tests and weightlessness on the contrary showed rather strong effects. The BN increased when the subjects were placed in a position with the last operated side under. Alternating linear accelerations provoked by pa-

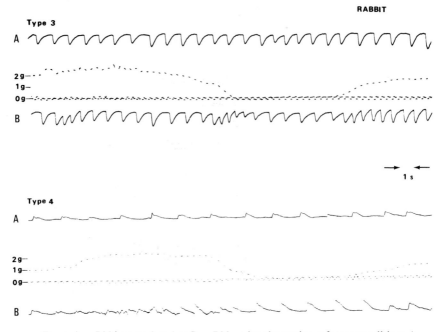

Fig. 1. A = BN in type 3 at 1 g; B = BN under alternating g forces condition; A = BN in type 4 at 1 g; B = BN during alternating g forces.

rabolic flight showed an effect on the BN according to four different types: type I: increase of amplitudes by high G values, decreasing in zero G; type II: increase of frequencies and decrease of amplitudes by high G values, decrease in zero G; type III: a slight effect on frequencies by alternating G forces; type IV: increase of amplitudes in zero G (fig. 1). When the BN was 3 days old in pigeons, it inverted during zero G (fig. 2).

Discussion

Magnus [6] found BN even after removal of the cerebrum or section of the upper cervical roots. The medulla oblongata solely proved to be sufficient. *Spiegel and Démétriades* [9] still found BN after removal of cerebellum, lesions of corpora quadrigemina or destruction of the vestibular nucleii on the side last operated upon. Destruction of the

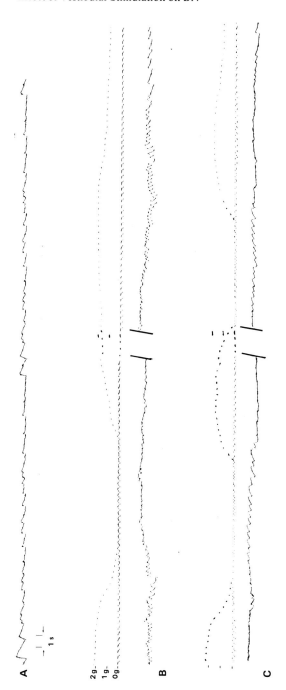

Fig. 2. A = BN under 1 g condition; B = BN during parabolic flight *1* day after second labyrinthectomy; C = inverted BN during weightless phase of parabolic flight 3 days after second labyrinthectomy.

vestibular nucleii on the other side eliminates BN [10]. This suggests that the activity of the vestibular nucleii without afferents are 'regenerated' and then counterbalance the activity of the opposite side. According to the 'defect theory' [1, 2], the symptoms of labyrinthine lesions are due to a loss of the resting activity and are partially compensated by the other sense organs. Others [3, 8] who assume that labyrinthine and posture receptors are automatically active support this. The effect of linear accelerations on the BN can be due to an input of the somatosensory system into the oculomotor system. This explains the inversion of BN in pigeons 3 days after the second labyrinthectomy as a temporary stage of a neuroplastic compensation process [7]. The fact that in spite of carefully sustained stimulus conditions more than 50 'Bechterew-compensated' animals expose an inconstant nystagmus, supports this assumption.

References

1   Bechterew, W.: Pflügers Arch. ges. Physiol. *30:* 312–347 (1883).
2   Goltz, F.: Pflügers Arch. ges. Physiol. *3:* 172–193 (1870).
3   Holst, E. von: Z. vergl. Physiol. *32:* 60 (1950).
4   Janeke, J.B.: On nystagmus and otoliths; thesis, Amsterdam (1968).
5   Kleyn, A. de: Graefes Arch. Ophthal. *107:* 408 (1922).
6   Magnus, R.: Körperstellung, p.352 (Springer, Berlin 1922/1924).
7   Schaefer, K.P.; Meyer, D.L.: In Zippel, Neurology and transfer of information, pp. 203–232 (Plenum Press, New York 1973).
8   Schöne, H.: Z. vergl. Physiol. *45:* 57–87 (1962).
9   Spiegel, E.A.; Démétriades, T.D.: Pflügers Arch. ges. Physiol. *210:* 215–222 (1925).
10 Uemura, T.; Cohen, B.: Prog. Brain Res. *37:* 515–527 (1972).
11 Versteegh, L.: Acta oto-lar., Stockh. *11:* 393 (1927).

Prof. Dr. W. J. Oosterveld, Academisch Medisch Centrum, K.N.O.-Vestibulair onderzoek, Meibergdreef 9, NL–1105 AZ Amsterdam (The Netherlands)

Medical Treatment of Vertigo.
How to Evaluate Its Effect?

Adv. Oto-Rhino-Laryng., vol. 30, pp. 345–349 (Karger, Basel 1983)

Panel Discussion Synthesis:
How to Evaluate the Effect of
Medical Treatment of Vertigo

Moderator: Makoto Igarashi[a]
Panelists: W.J. Oosterveld, Jens Thomsen, Isamu Watanabe,
Wallace Rubin

[a] Department of Otorhinolaryngology, Baylor College of Medicine,
Texas Medical Center, Houston, Tex., USA

Selection of Drug

In order to select a drug for a patient with vertigo, first the patient needs to be thoroughly worked up. The choice of the drug must be based upon whether the drug will be used to treat vertigo etiologically, or to relieve this unpleasant complaint. The site of pharmacological action therefore must be contemplated even though the effects of drugs are in many cases widely distributed and not selectively to a single specific area.

It is important to recognize how the clinical figures of vertigo change from time to time among different patients. Thus, it is necessary to classify and determine what kind of and what stage of vertigo we are dealing with. The choice of treatment should be different in accordance with the different stages. When the drug is used for the control of acute episodes, many drugs, such as a vestibular depressant, will suppress the symptoms almost immediately and effectively. However, a pharmacological dilemma exists in that the drug effectively controlling the subjective ill-feeling could be delaying the equilibrium system compensation or recovery. If the vertigo is the non-persistent periodical type, a prophylactic medication, such as some kind of diuretic, can be used during the intermittent stage. Often the differentiation of peripheral or central causes of vertigo is not clear, but it will definitely assist the choice of drug. The method of administration and the side effects of the drug must be always taken into consideration. The evaluation done at different times after administration could show different results depending upon the mode of drug concentration and retention.

Manner of Evaluation

Treating vertigo is not the same as treating some clinically definable disease, such as tumor. Treating vertigo is treating the patient as a whole, and many parameters must be evaluated. Because vertigo is a very subjective feeling, some uncertainty may remain in diagnosis. Nevertheless, treatment is very much needed.

Even though in many drugs the explanation of the effect on vertigo is somewhat hypothetical, the evaluation must be based on recording daily condition of the vertigo. Usually, no objective parameters are available to judge the effect of the drug on vertigo.

The assessment of drug efficacy is not easy, basically due to a tremendous variability of this symptom. Variability could be induced by many indirect factors which include: diet and metabolic factors (water, salt, sugar, etc.), smoking and drinking, allergy, immune reaction, hormone, autonomic nervous system, vitamins, infections, poisoning, heredity, etc. Furthermore, background or outside factors, such as psychological stress, vocational situation, life style, love affairs, family problems, living environment, climate, etc. cannot be ignored.

For the strict evaluation of drug efficacy, the aforementioned extra factors and compatible treatments other than drugs need to be controlled, otherwise these may bias the investigational results. However, usually vertiginous patients are very susceptible to whatever is involved; therefore, the control of these extras in reality may be difficult and sometimes even not advisable. If all extras could not be strictly controlled during the investigation, a global evaluation in a clinical trial (even though it lacks a good quantification) may be able to provide similar information.

The most important thing is not the objective test findings, but the patient's opinion about his symptoms. However, tests may provide less biased biomedical responses, whereas the evaluation of the patient's opinion can be biased by the manner of questioning, psychological effect, intellectual level, etc.

The data from vestibulometry and audiometry may not necessarily reveal the same results compared to the evaluation of subjective symptoms. A difference in precision could be one factor to produce this discrepancy. Regardless of this discrepancy, if the drug is effective to relieve subjective vertigo, it is effective. Another possible explanation of this discrepancy is the fact that these tests are usually done only at a certain time which may not be well representing the overall condition of vertigo that can have a fluctuating course. For this reason, frequent monitoring of the objective signs, such as nystagmus (recorded by a portable recorder whenever the recording is needed), could be profitable in addition to the patient's opinion on the subjective symptoms.

Evaluation of drug efficacy in provoked nystagmus or motion sickness has its own value, even though many limitations exist; however, the results cannot be directly transferable for the treatment of vertigo.

Investigational Design

The controlled trials to evaluate drug efficacy have many practical weaknesses. It is nevertheless indispensable to make a scientifically sound determination of whether a given treatment is indeed suitable for use in vertigo.

The detailed planning is most important and the more the investigator has anticipated before the trial begins, the more likely are the results to be meaningful. It is necessary that all individuals who take part in the trial must follow the exact plan, for any variation will reduce the quality of the investigation. A standardized method of dealing with the patient is necessary. The questions and the explanations regarding the trial and the anticipated effect of the drug, must be identical for all participants.

In the planning stage, special attention must be paid to the fact that vertigo could be influenced by a number of 'difficult-to-control' non-specific factors. Also, symptom variability among the unselected group of vertiginous patients reduces the value of investigation. Therefore, it is necessary before the study begins to develop a strict definition of the patient selecting manner. However, if the sample number becomes too small, it may be critical in order to make a statistical evaluation valid.

For an experimental design, it is important to define whether the aim of the study is to question if the drug has indeed any effect on vertigo, or to determine whether the drug is more effective compared to the already available one. In the former case, it is necessary to have a placebo; however, the placebo effect is also variable and it often masks the drug effect. It could be advantageous to demonstrate the correlation between different dose levels of the drug on trial, but not for the placebo.

Because only the patient's own judgement is available for the evaluation of the drugs, it is necessary to have thorough information which should include vertiginous duration, severity, frequency, associated symptoms, working capacity, etc. An importance exists to closely follow-up the patient under the investigator's care. By doing it this way, even a placebo can be effective.

As an investigational design, a strict double-blind technique is necessary. The use of randomized peer groups, regarding age, sex, duration, and severity of illness, etc., could be most desirable. The advantage of the cross-over method is that inter-individual variations disappear and the treatment and the control groups become identical. However, the disadvantage is that the spontaneously changing disease course may introduce a variable along the longitudinal time course. Another contamination will be the prolonged or cumulative effect of a used drug after the withdrawal; therefore, enough length of pause is needed. The cross-over trial can be carried out for more than two periods.

Placebo and Medico-Ethical Problems

Different opinions exist in ethical viewpoints about the double-blind placebo study, even though it is generally accepted that placebo is a sort of psychological treatment which should have a profitable effect on vertigo. Placebo is not equal to entire null treatment.

A difference exists between different countries in regard to medico-ethical, medico-legal and medico-sociological aspects. In some countries, the subject patient must be fully explained about the possibility that he may receive a placebo, whereas in other countries if the rationale why the investigator does not want to give the patient the full explanation is convincingly strong enough, placebo studies may be permissible without giving the patient full detailed information. The Helsinki Declaration may cause a severe drop-out problem, and may make the study with only a limited patient group.

Prof. M. Igarashi, Department of Otorhinolaryngology, Baylor College of Medicine, Otological Research Laboratory, Texas Medical Center, Houston, TX 77030 (USA)

Adv. Oto-Rhino-Laryng., vol. 30, pp. 350–354 (Karger, Basel 1983)

Ménière's Disease: A 3-Year Follow-Up of Patients in a Double-Blind Placebo-Controlled Study on Endolymphatic Sac Shunt Surgery

J. Thomsen, P. Bretlau, M. Tos, N.J. Johnsen

University ENT Department, Gentofte Hospital, Hellerup, Denmark

In 1981, we published the results of a double-blind placebo controlled study on endolymphatic sac shunt surgery [*Thomsen* et al., 1981]. The results were based upon daily ratings of all the symptoms within the Ménière's syndrome: nausea, vertigo, hearing impairment, tinnitus and pressure in the ear for 1 year after either a regular, Silastic sheath shunt draining into the mastoid cavity, or a simple mastoidectomy. Minor differences could be established between active and placebo treatment, but the greatest differences in symptoms were found when comparing preoperative and postoperative scores, where both groups improved significantly. 15 patients had the active surgery and 15 the placebo surgery.

In this paper we present a 3-year follow-up study of the same patients.

Methods and Patients

Of the original 30 patients with typical Ménière's disease, refractory to medical treatment, 4 patients eluded the 3-year follow-up. 3 patients have died and 1 patient has moved to Spain and cannot be reached. Fortunately, these drop-outs do not alter the material significantly since two deaths occurred in the placebo group, while 1 who died and the remaining lost patient had active treatment.

The patients are still unaware of which kind of surgery was performed, and the investigator who has controlled the patients and made the interviews still does not know which groups the patients belong to.

Table I. AAOO reporting criteria

| AAOO | 1 year | | | | 3 years | | | |
| | active | | placebo | | active | | placebo | |
	n	%	n	%	n	%	n	%
A	1	7	1	7	3	23	2	15
B	12	80	6	40	6	46	8	62
C	0		3	20	0		2	15
D	2	13	5	33	4	31	1	8
Total	15		15		13		13	

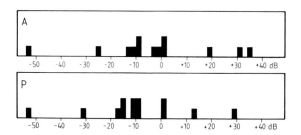

Fig. 1. Histogram showing the intrasubjective difference in dB in the preoperative and postoperative (36 months) audiogram (median values) for hearing loss at 250, 500 and 1000 Hz in active (A) and placebo (P) group.

Results

The AAOO classification for the patients is seen in table I. At 3 years there is no difference between the distribution in the classes when active is compared to placebo. Our usual indication for success (classes A, B and C) is absence of vertigo, and 4 patients in the active group still had periodic attacks, while only 1 patient in the placebo group claimed to have vertigo. If only classes A and B are included in the success group no difference can be detected between the groups. In analyzing the intrasubjective difference between the median values of the preoperative and postoperative (36 months) objective parameters of each fre-

Table II. Patients evaluation of operative effect

Operation	Good		Reasonable		Poor		Total	
	n	%	n	%	n	%	n	%
1 year								
Active	11	73	3	20	1	7	15	100
Placebo	10	67	5	33	0	0	15	100
Total	21		8		1		30	100
3 years								
Active	9	70	2	15	2	15	13	100
Placebo	10	77	3	23	0	0	13	100
Total	19		5		2		26	100

Table III. Investigators' evaluation of operative effect

Operation	Good		Reasonable		Poor		Total	
	n	%	n	%	n	%	n	%
1 year								
Active	11	73	2	13.5	2	13.5	15	100
Placebo	12	80	3	20	0	0	15	100
Total	23		5		2		30	100
3 years								
Active	9	69	1	8	3	23	13	100
Placebo	10	77	2	15	1	8	13	100
Total	19		3		4		26	100

quency (250, 500, 1000, 2000 and 4000 Hz), no statistical differences could be shown at the 5% level (fig. 1).

Table II shows the patients' evaluation of the operative effect. 70% in the active group found good benefit from the operation compared to 77% in the placebo group. Only 2 patients found a poor effect of surgery and both patients had had the active shunt inserted.

The investigators' evaluation of the operative effect is shown in table III, where 69% in the active shunt group were described as good, against 77% in the placebo group. This difference is not significant, and definitely not in favour of the shunt. 3 patients with shunt were described as failures, while only 1 of the patients in the placebo group could be placed in the failure group.

Discussion

Our paper [*Thomsen* et al., 1981] gave rise to some criticism about the results of the study [*Arenberg*, 1981; *Vaisrub*, 1981] as well as of the study design. Granted, the 15 patients in each group is a rather small number, but the small sample number has of course been taken into account in the statistical tests. We have not buried significant changes under insignificant averages, as suggested by *Arenberg* [1981] since each patient has been used as his own control, and averaging of data has not occurred. Each frequency (250, 500, 1000, 2000 and 4000 Hz) has been tested separately, and only at 250 Hz could a statistical difference ($p < 0.05$) be demonstrated after 1 year. At the 3-year examination this difference could not be upheld.

Our 3-year results seem to confirm our results from the first year: *no* significant differences can be established between the effect upon the symptoms in patients with severe Ménière's disease having a Silastic sheath endolymphatic sac shunt, draining to the mastoid cavity, or having a simple mastoidectomy. About 70% in both groups could be classified as successes, with nearly complete alleviation of symptoms. We are aware of the fact that we can only speak for the Silastic sheath mastoid shunt, and that we cannot with certainty exclude that another procedure, e.g. Arenberg unidirectional valve, or any shunt draining to the subarachnoid space, might be superior. However, the high and persistent success in our placebo-treated patients tends to speak against this possibility.

As physicians, we have to offer the patients the best possible treatment. And this may well include some kind of surgery in patients with Ménière's disease. We must, however, be aware of the possibility of a very strong placebo effect, especially in these troubled patients, and not lead ourselves to believe the effect to be the result of a specific, e.g. shunt, procedure alone.

References

Arenberg, J.K.: Placebo effect for Ménière's disease sac shunt surgery disputed. Letter to the editor. Archs Otolar. *107:* 773 (1981).

Thomsen, J.; Bretlau, P.; Tos, M.; Johnsen, N.J.: Placebo effect in surgery for Ménière's disease. Archs Otolar. *107:* 271–277, 1981.

Vaisrub, N.: Letter to the editor. Archs Otolar. *107:* 773 (1981).

J. Thomsen, MD, University ENT Department, Gentofte Hospital, DK–2900 Hellerup (Denmark)

Adv. Oto-Rhino-Laryng., vol. 30, pp. 355–361 (Karger, Basel 1983)

Subjective and Objective Evaluation of Medical Treatment for Ménière's Disease, with Special Reference to the Dose Response for Adenosine Triphosphate

K. Mizukoshi[a], *Y. Watanabe*[a], *I. Watanabe*[b], *J. Okubo*[b],
To. Matsunaga[c], *Ta. Matsunaga*[d], *S. Takayasu*[e], *I. Kato*[f],
T. Tanaka[g]

[a] Toyama Medical and Pharmaceutical University, Toyama;
[b] Tokyo Medical and Dentistry University, Tokyo;
[c] Osaka University, Osaka; [d] Nara Medical University, Nara;
[e] National Hospital, Tachikawa; [f] Yamagata University, Yamagata;
[g] Tokyo University, Tokyo, Japan

Introduction

Vertigo is a subjective sensation, so it is very difficult to evaluate. In Ménière's disease, the severity of vertigo and/or dizziness is extremely variable and the degree of auditory and vestibular symptoms is frequently not parallel. Also, there are great variations in the clinical symptoms in the patients with peripheral vestibular disorders. Therefore, for each patient the treatment of Ménière's disease should be evaluated from the standpoint of the following three conditions: (1) the controlled effects, i.e., on the intensity and the duration of vertiginous attacks; (2) the suppressive effects, i.e., on the frequency of the attacks, and (3) the improvement effects, i.e., on such inner ear functions, as hearing and equilibrium.

In order to determine the maximum dose response of adenosine triphosphate (ATP), subjective and objective evaluations of its effects on Ménière's disease and other peripheral vestibular disorders were made in a double-blind control study comparing doses of 300 and 150 mg given daily for 4 weeks.

Table I. Matched pair groups of the control study

	ATP 300 mg	ATP 150 mg	Total
Male	27	31	58
Female	53	43	96
Age, years (average)	43.4 + 1.5	42.7 + 1.5	42.9 + 1.1
Ménière's disease	31	38	69
BPPV	10	8	18
Vestibular neurotonis	1	4	5
Others	38	24	62
Total	80	74	154

Subjects and Methods

From ten ENT departments of University Hospitals, 154 patients with peripheral vestibular disorders were selected for this study. Details of patients' history and distribution are shown in table I. The diagnosis was based on the criteria of the Ménière's Disease Research Committee of Japan [1].

There was no significant difference between the two groups (300 and 150 mg daily doses ATP) as to age, sex, or disease distribution. The clinical subjective and objective symptoms in the two groups were assessed every 2 weeks both by the doctors and by the patients themselves. Objective symptoms per the equilibrium and auditory examinations were evaluated at the first examination before prescribing treatment, again after 2 weeks, and finally after 4 weeks, always by the same doctors. 300 and 150 mg daily doses of ATP were given at random in individually coded opaque capsules of identical appearance.

Results

Subjective Evaluation of Symptoms

Subjective symptoms (table II) were evaluated every two weeks, for intensity, duration and frequency of the vertiginous attacks. Moreover, the associated symptoms, such as tinnitus, hearing impairment, fullness of the ear, nausea/vomiting, and headache were assessed. The extent of improvement in the subjective symptoms was graded using the following scale of five degrees: (1) marked improvement; (2) moderate improvement; (3) slight improvement; (4) no change, and (5) some deterioration. Results were compared with preceding evalua-

Table II. Global judgement of subjective symptoms

Symptoms	Rate of improvement		Statistical significance
	ATP 300 mg	ATP 150 mg	
Vertigo/dizziness			
Vertigo	88.6	80.6	n.s.
Dizziness	82.3	70.0	n.s.
Block-out	60.0	50.0	n.s.
Associated symptoms			
Tinnitus	56.3	61.0	n.s.
Hearingloss	28.3	25.6	n.s.
Fullness of the ear	52.8	43.3	n.s.
Nausea, vomiting	87.0	70.7	n.s.
Headache	*76.5*	*52.0*	$p < 0.1$

n.s. = No significance. Percent values.

Table III. Global judgement of objective symptoms

Tests	Rate of improvement		Statistical significance
	ATP 300 mg	ATP 150 mg	
Mann's test	67.7	64.7	n.s.
Stepping test	68.3	75.0	n.s.
Spontaneous nystagmus	*77.5*	*56.4*	$p < 0.05$
Positional nystagmus	*81.4*	*66.7*	$p < 0.05$
Positioning nystagmus	*70.6*	*53.8*	$p < 0.1$
Caloric nystagmus	*81.8*	*39.1*	$p < 0.01$
Hearing loss	24.1	34.0	n.s.

Percent values.

Table IV. Global judgement of clinical evaluation

Dosage	Improved	Not improved	Statistical significance
ATP 300 mg	68 (89.5%)	8	$p < 0.05$
ATP 150 mg	49 (75.4%)	16	

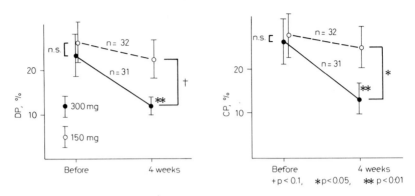

Fig. 1. Effects of 300 and 150 mg ATP daily upon the caloric nystagmus.

tions. For each patient in the two groups, a modification of the symptoms was noted statistically analyzed. As shown in table II, with the exception of headache, there was no significant difference between the two groups revealed in the subjective symptoms.

Objective Evaluation of Symptoms

The objective evaluation (table III) included examination of both auditory and equilibrium functions. Investigation of spontaneous, positional, and caloric nystagmus was made by using electronystagmography (ENG). In addition, Mann's test and Fukuda's stepping test were applied in this study. However, auditory function was only evaluated by pure-tone audiometry during this control trial.

A statistical comparison of the objective symptoms of the two dosage groups revealed a significant difference in spontaneous, positional and caloric nystagmus test results. However, as to hearing improvement no significant different preference between the two dosages was concluded from this study (table III).

Further, in the caloric nystagmus tests, the directional preponderance (DP) and the canal paresis (CP) were calculated in a percentage manner, according to the formula already reported by *Jongkees and Philipszoon* [9]. Here the duration of the caloric responses was used as a parameter of this analysis. As represented in figure 1, after the control trials, the duration percentages of the DP and CP of the 300-mg group showed more significant improvement than did those of the 150-mg group.

From these subjective and objective evaluations of the dose response of ATP, the global judgement of the attending doctors showed that the 300-mg ATP dosage was significantly more preferable than the 150-mg dosage when the subjective and objective global evaluations of the patients with peripheral vestibular disorders were statistically analyzed by the matched pair group method. This is represented in table IV. Side effects were minimal, and only 4 patients reported tinnitus, sleepiness, stomatitis and constipation.

Discussion

It is generally accepted that ATP markedly improves cerebral blood flow and inner ear metabolism, and suppresses vestibular responses [2–5]. Therefore, the antivertiginous effects of ATP should be evaluated from the standpoint of these effects in patients with peripheral vestibular disorders.

In this respect, we have studied the different dose response of 154 patients with the peripheral vestibular disorders clinically evaluating their subjective and objective symptoms. From these clinical studies, no significant difference between the effects of 300 and 150 mg daily doses of ATP on subjective symptoms was recognized for controlling any of the clinical symptoms except headache. However, the evaluation of the effect of the different dosages on objective symptoms showed a more significant difference between the effects of the two dosages on spontaneous, positional and caloric nystagmus test results in the double-blind control study.

An analysis of our caloric nystagmus tests seems to indicate a correlation between the dosage of ATP and the suppressive effects on the vestibular nystagmus responses, using the duration as the parameter of analysis. This suppressive effect on vestibular responses has already been reported by *Nakamura* et al. [6] by using postrotatory nystagmus.

In order to determine the maximum dose response for antivertiginous drugs, it is recommended that any statistical comparison between the effects of different dosages on vertiginous attacks should be made by the matched pair group method. For this method, many controlled patients must be collected, and the objective and subjective evaluations of the antivertiginous effects should be done in clinical practice.

In this respect, the AAOO criteria for reporting results of any therapeutic evaluation in Ménière's disease and other inner ear disorders have been extremely helpful [7]. However, there is great variation in the stage of inner ear disease. Therefore, various investigators have reported evaluation of the medical treatment of Ménière's disease. Such a wide variety of clinical evaluations suggests that the staging system for Ménière's disease and/or any inner ear disorder is a very helpful method as reported by *Arenberg and Stahle* [8]. However, it could be argued that 4 weeks of treatment with ATP for inner ear disorders is not sufficiently long enough to evaluate the effects on inner ear disorders. From our double-blind controlled study, it should also be emphasized that any statistical comparison of the dosages of antivertiginous drugs should be analyzed by the matched pair group method as this method is the preferable method for clinical evaluation of the maximum dose response in patients with vertigo and/or dizziness. However, further study is needed to determine what the most appropriate system method for evaluating the effects of medical treatment of inner ear disorders is.

Acknowledgements

Grateful acknowledgement is made to the staff members of this controlled study for their kind assistance and for providing their clinical results.

References

1   Watanabe, I.: Ménière's disease Research Committee of Japan. Proc. 5th Extraord. Meet. Bárány Soc., pp. 281–283, Int. J. equilib. Res., suppl. (1975).

2   Gottsein, U.; Niedermayer, W.: Tierexperimentelle Untersuchungen über die Wirkung von Adenosin-mono- und Adenosin-triphosphat auf die Hirndurchblutung. Klin. Wschr. *36:* 972–975 (1958).

3   Koide, Y.; Sasaki, S.; Nakano, Y.; Nagashima, N.: Some aspects of medical treatment of inner ear disease. Acta med. biol. *8:* 295–236 (1961).

4   Gotoh, K.; Muranushi, Y.: Treatment for the positioning nystagmus. With special reference to ATP. Otologia, Tokyo *39:* 879–888 (1967).

5   Jakobi, H.; Spinar, H.; Kuhl, K.D.; Lotz, P.; Haberland, E.J.: ATP-Anwendung bei Innenohrerkrankungen in der Klinik und im Experiment. Acta oto-lar. *83:* 195–199 (1977).

6   Nakamura, M.; Yokoyama, T.; Shirasawa, T.; Hirai, K.; Sano, N.: Effects of adenosine triphosphate on the postrotatory nystagmus and disorders of vestibular function. Folia pharmacol. jap. *75:* 487–494 (1979).

7   Alford, B.R.: Report of subcommittee on equilibrium and its measurement. Trans. Am. Acad. Ophthal. Otolar. *76:* 1462–1464 (1972).
8   Arenberg, I.K.; Stahle, J.: Part IV. Staging the aural aspects of Ménière's disease. Endolymphatic sac valve implant surgery. Laryngoscope *89:* suppl. 17, pp. 40–47 (1979).
9   Jongkees, L.B.W.; Philipszoon, A.J.: The caloric test in Menière's disease. Acta otolar., suppl. 192, pp. 168–170 (1964).

K. Mizukoshi, MD, Department of Otolaryngology, Toyama Medical and Pharmaceutical University, Toyama 930-01 (Japan)

Adv. Oto-Rhino-Laryng., vol. 30, pp. 362–364 (Karger, Basel 1983)

A Critical Study on the Evaluation of the Effect of Treatment of Ménière's Disease

Hitoshi Kitano, Masaaki Kitahara

Department of Otolaryngology, Shiga University of Medical Science, Seta, Otsu, Japan

Introduction

In Ménière's disease, there are numerous medical and surgical treatments for vertigo. However, as the vertigo with Ménière's disease is characterized by irregular remission, it is difficult to evaluate the effect of treatment.

In this study, we have made a critical evaluation concerning two criteria for evaluating the effect of treatments which have been used worldwide. We then propose a reasonable criterion to evaluate medical treatment.

Long-Term Results

Concerning evaluation of the effect of treatment, many researchers have tried to evaluate how vertiginous attacks are controlled during a definite follow-up period. Most of them hope for long-term results and actually have applied long observation periods of from 2 to 5 years to evaluate whether the treatment is effective.

A survey was conducted of medical students who were asked what they would want to know if they were a patient who intended to receive surgery for Ménière's disease. The given choices were three parameters based on long-term results, average interval of attacks and maximum interval of attacks, respectively. 59 out of 83 (71%) medical students replied that they would want to be assured of the probability that no vertiginous attack would occur for at least 5 years after surgery. From the above-mentioned results, we found that this evaluation of treatment was helpful to enable patients to decide whether to receive the treatment or not. However, the problem with this criterion is that it is impossible to obtain results of the treatment before 5 years have passed. Furthermore, even if we have the results for a 5-year follow-up period, we cannot evaluate whether the treatment is effective without a control for the comparison of evaluation.

Average Interval of Attacks

This is a criterion for evaluation of the effect of treatment which was made by the AAOO in 1972. Vertiginous control of AAOO refers to the absence of definitive attacks for 10 times of the average interval between attacks before treatment.

This criterion, however, posed many difficulties when applied strictly. First, the patient's history is usually restricted to what he can recall and it is difficult to remember exactly the entire history of vertiginous attacks. Second, even if we can obtain the accurate average interval of attacks, long-term observation is needed in the case of long-term average interval of attacks.

Maximum Interval of Attacks

As the above-mentioned criteria posed some difficulties for practical use, we propose the following criterion by means of maximum attack-free interval.

We put together the cumulative frequency polygon of maximum attack-free intervals covering the period from the first vertiginous attack to the first visit to our clinic in which 61 patients with Ménière's disease did not have any special treatment and were collected at random.

By using this polygon, we can presuppose a rate of frequency of patients in which vertigo may occur. For instance, in patients with an attack-free interval of 1 year, we found that in 70% patients with Ménière's disease, there was a possibility of occurrence of a vertiginous attack afterwards. On the contrary, we found that 30% patients with Ménière's disease have the maximum attack-free interval within a 1-year period. Furthermore, it is possible to evaluate the effectiveness of any special treatment at any follow-up period. For instance, when a rate of the cumulative frequency of patients with Ménière's disease who received a special treatment is significantly more than 20% during the 5-year follow-up period, it provides evidence of its effect.

It is also confirmed that the duration of illness has little influence on the cumulative frequency polygon. We examined the results of epidural shunt operation with Ménière's disease by means of this criterion.

32 out of 39 patients (82%) had no vertiginous attack during the 5-year follow-up period after the operation. Comparing the results of operation with the polygon, we can find out the effectiveness of the operation.

Though the criteria used worldwide have some problems, including the difficulty in obtaining the history of illness and the necessity of long-term observation, it is rather easy to obtain the maximum attack-free interval from the patient's history. Furthermore, we can evaluate the effectiveness of treatment at any follow-up period by using the cumulative frequency polygon of maximum attack-free interval. The polygon is considered to be a practical method for evaluating the effect of treatment.

H. Kitano, MD, Department of Otolaryngology,
Shiga University of Medical Science, Seta, Otsu 520-21 (Japan)

Adv. Oto-Rhino-Laryng., vol. 30, pp. 365–369 (Karger, Basel 1983)

Scopoderm-TTS® (Scopolamine) Influence on Caloric-Induced Nystagmus: an Extract

S. B. Larsen, E. Peitersen

University ENT Department, Hvidovre Hospital, Copenhagen, Denmark

A double-blind, cross-over study in which 14 healthy volunteers (7 female and 7 male) in randomized order received single or double dose levels of Scopoderm-TTS® and placebo in order to evaluate the efficacy of Scopoderm-TTS® on the vestibular function, and to reveal the extent of well-known side effects. The effect on the vestibular function can be demonstrated by a suppression of the caloric-induced nystagmus, maximum eye speed (MES) of the slow phase as well as the duration of nystagmus.

Scopoderm-TTS® is a new medical drug primarily against motion sickness, consisting of a reservoir of scopolamine contained in a four-coated plaster film, a so-called transdermal therapeutic system (TTS), to be released in an exact dose, i.e. 5 $\mu g/h$, after being applied on the postauricular area. Clinical tests in the USA have shown that skin-applicated scopolamine gives a constant plasma concentration in steady state as well as continuous intravenous infusion.

In this study the three parameters: MES, the duration, and the sensation of rotation were examined after caloric-induced nystagmus.

Results

Figure 1 shows the percentage reduction in MES in proportion to the basic value by 1 and 2 TTS-Scopoderm and placebo respectively. Both placebo and 1 and 2 TTS-Scopoderm give a significant percentage reduction in MES. The percentage reduction in *duration* (fig. 2) is also significant in placebo as well as in 1 and 2 TTS-Scopoderm, but

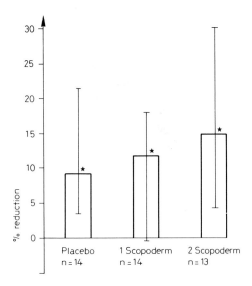

Fig. 1. Percentage reduction in MES (median ± 95% confidence interval).

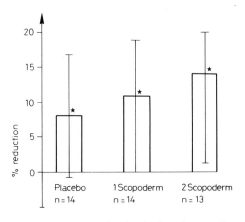

Fig. 2. Percentage reduction in duration (median ± 95% confidence interval).

with a significantly higher effect of 2 TTS-Scopoderm than of 1 TTS-Scopoderm in comparison with placebo.

Sensation of rotation was graphically registered on a visual analogue scale. We found a significant effect of both 1 and 2 doses of TTS-Scopoderm but *no* significant effect of placebo (fig. 3).

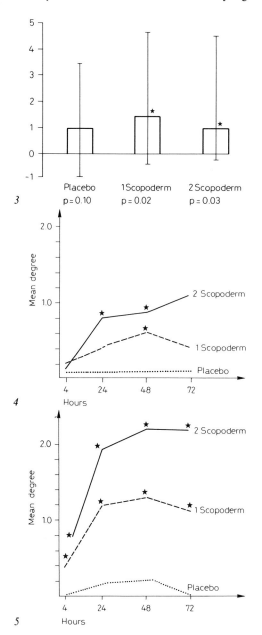

Fig. 3. Reduction in sensation of rotation (cm, median ± 95% confidence interval).
Fig. 4. Bluntness.
Fig. 5. Dryness of the mouth.

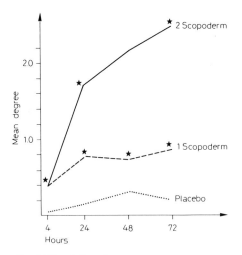

Fig. 6. Visual disturbances.

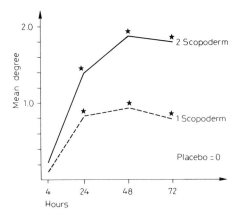

Fig. 7. Other side effects.

Regarding *side-effects* we asked specifically about *bluntness* (fig. 4), *dryness of the mouth* (fig. 5), *visual disturbances* (fig. 6) and *other side effects* all together (fig. 7). Side effects that may arise had to be scored on a scale from 0 to 3, respectively 4, 24, 48 and 72 h after application.

Figure 4–7 illustrate graphically a clear significant and dose-related difference between placebo and 1 and 2 TTS-Scopoderm, and apart from bluntness a significant difference between single and double dose of Scopoderm. Furthermore, we found a higher degree of side effects in women than in men. However, in the therapeutic dose, that is 1 TTS-Scopoderm, in *no* individual case were side effects registered exceeding 'slightly uncomfortable'.

Conclusively Scopoderm-TTS® in a single dose seems to be an effective and useful vestibular depressant preparation, while administration in larger doses seems unadvisable and should only take place on specific indications.

S.B. Larsen, MD, University ENT Department, Hvidovre Hospital,
Copenhagen, DK-2650 Hvidovre (Denmark)